An Introduction to Time

Dr.A.K.Saxena, Ph.D.

Published by Dr.A.K.Saxena, Ph.D., 2024.

While every precaution has been taken in the preparation of this book, the publisher assumes no responsibility for errors or omissions, or for damages resulting from the use of the information contained herein.

AN INTRODUCTION TO TIME

First edition. November 10, 2024.

Copyright © 2024 Dr.A.K.Saxena, Ph.D..

ISBN: 979-8227236654

Written by Dr.A.K.Saxena, Ph.D..

Also by Dr.A.K.Saxena, Ph.D.

Handbook for Introverted Leaders: Strategies for Success
No Longer a Yes Man
Fasting, Feasting and Spirituality
The "Karma" Puzzle
Open Secrets
15 Things You Should not Worry About
Why Denial of Death?
From Sorrow to Serenity
Unlock Your Inner Power and Potential
An Introduction to Time

Table of Contents

PREFACE ... 1
Dedication ... 3
CHAPTER 1 | INTRODUCTION ... 4
CHAPTER 2 ... 7
CHAPTER 3 ... 10
CHAPTER 4 ... 14
CHAPTER 5 ... 17
CHAPTER 6 ... 20
CHAPTER 7 ... 24
CHAPTER 8 ... 28
CHAPTER 9 ... 31
CHAPTER 10 ... 34
CHAPTER 11 ... 37
CHAPTER 12 ... 40
CHAPTER 13 ... 44
CHAPTER 14 ... 48
CHAPTER 15 ... 52
CHAPTER 16 ... 55
CHAPTER 17 ... 58
CHAPTER 18 ... 61
CHAPTER 19 ... 64
CHAPTER 21 ... 67
CHAPTER 21 ... 70
CHAPTER 22 ... 73
CHAPTER 23 ... 76
CHAPTER 24 ... 79
CHAPTER 25 ... 82
CHAPTER 26 ... 85
CHAPTER 27 ... 91

Time is a keyhole.... We sometimes bend and peer through it. And the wind we feel on our cheeks when we do--the wind that blows through the keyhole--is the breath of all the living universe.

-- Stephen King

PREFACE

"Time isn't precious at all, because it is an illusion. What you perceive as precious is not time but the one point that is out of time: the Now. That is precious indeed. The more you are focused on time—past and future—the more you miss the Now, the most precious thing there is."

— *Eckhart Tolle*, The Power of Now: A Guide to Spiritual Enlightenment

Time, the elusive fabric of our existence, weaves together the tapestry of human experience. We live within its rhythms, bound by its constraints, and yet, we struggle to grasp its essence. *"An Introduction to Time: Universe's Greatest Secret"* invites you on a fascinating journey to unravel the mysteries of time, exploring its depths, complexities, and profound implications.

For centuries, philosophers, scientists, and thinkers have pondered the nature of time. From ancient civilizations to modern theories, our understanding has evolved, but the enigma remains. This book aims to demystify time, rendering its complexities accessible to all, and sparking a deeper appreciation for the intricate dance between time, space, and human existence.

Within these pages, we will embark on an extraordinary exploration, navigating various aspects of time through clear explanations, engaging examples, and thought-provoking questions, this book will guide you in:

Recognizing time's impact on our daily lives
Comprehending time's relationship with space and gravity
Exploring the intersection of time and human consciousness
Pondering the implications of time's mysteries on our existence
Developing a deeper appreciation for the complexities of time

"An Introduction to Time" is not just a book about time; it is an odyssey through the human experience. As we navigate the intricacies

of time, we will discover new perspectives on our place within the universe and uncover the hidden patterns that govern our lives.

Join me on this captivating journey, as we unravel the threads of time and reveal the secrets hidden within.

Dedication

To all those who have ever pondered the mysteries of universe and of time.

Acknowledgments

Gratitude to:

The visionary thinkers, scientists, and philosophers who have illuminated our understanding of time.

The researchers and experts who have contributed to our modern comprehension of time.

My readers who have inspired me to share this fascinating story with them.

@@@@

CHAPTER 1
INTRODUCTION

Time has always been a fascinating and elusive concept that has puzzled philosophers, scientists, and theologians for centuries. It is an essential aspect of our daily lives, yet it is true nature remains a mystery. From ancient Greek philosophers like *Aristotle* to modern-day physicists like *Albert Einstein* and *Stephen Hawking*, the understanding of time has evolved, and with it, the paradoxes surrounding it have only deepened. In this book, *"An Introduction to Time: Universe's Greatest Secret"*, I discuss this complex and thought-provoking topic, exploring the various theories and perspectives on time.

The concept of time is deeply ingrained in our perception of reality. We measure our lives in seconds, minutes, and hours, and we use time as a tool to organize our daily activities. However, upon closer inspection, we realize that time is not as straightforward as it seems. As we go about our lives, we experience time as a linear progression from past to present to future. But is this how time truly functions?

Is time just an illusion created by our minds?

Aristotle, one of the most influential philosophers in history, believed that time was an objective reality that exists independently of human consciousness. He saw time as a continuous flow that could be divided into smaller units for practical purposes. For Aristotle, time was closely related to change and motion. He believed that the past and the future were merely potentialities, and only the present moment truly existed.

On the other hand, the ancient Greek philosopher *Parmenides* argued that time was an illusion and that only the present moment existed. He believed that our perception of past and future was a result of our limited understanding, and that true reality was timeless and unchanging. This idea challenges our common understanding of time

AN INTRODUCTION TO TIME

as a linear progression and raises questions about the nature of existence itself.

The concept of time continued to intrigue philosophers and scientists throughout history, and in the 17th century, *Isaac Newton* presented his groundbreaking theory of absolute time. According to *Newton*, time was an absolute and universal framework that existed independently of any object or observer. He saw time as a constant flow that was the same for everyone, regardless of their location or perspective. This idea had a significant impact on the scientific world, providing a precise measure of time that could be used for calculations and experiments.

However, in the early 20th century, *Albert Einstein's* Theory of Relativity challenged Newton's concept of absolute time. *Einstein* proposed that time and space were interwoven and that they could be distorted by gravity and motion. He introduced the concepts of space-time and time dilation, which showed that time could pass at different rates for different observers depending on their relative speed. This idea revolutionized our understanding of time and paved the way for further research and exploration into the nature of this mysterious concept.

One of the most intriguing paradoxes surrounding time is the twin paradox, which is based on Einstein's theory of relativity. It states that if one twin travels at high speeds in space while the other stays on Earth, the traveling twin will age slower than the stationary twin. When they reunite, the traveling twin will have experienced less time than their sibling, creating a paradox where both twins have aged differently. This paradox highlights the subjective nature of time and challenges our understanding of it as a linear and objective concept.

As we look deeper into the paradoxes of time, we are forced to question what exactly is true and real. Is it our perception of an ever-flowing river of time, or is it just an illusion created by our minds? And if it is an illusion, then what is the true nature of reality?

In my book, *"An Introduction to Time: Universe's Greatest Secret"* I have tried to explore time and the implications in detail. I take readers on a journey through the history of time, from ancient Greece to modern-day theories and research. I discuss the philosophical, scientific, and psychological aspects of time, examining how our perception of it has shaped our understanding of the world and universe.

Whether you are a philosopher, scientist, or simply someone intrigued by the mysteries of time, this book is for you. Through research and analysis, I aim to shed light on the paradox of time and provide readers with a comprehensive and detailed understanding of this complex concept. We will explore the evolution of our understanding of time and the implications it has on our perception of reality of universe.

In conclusion, the concept of time remains a mystery that continues to fascinate and challenge us. Throughout history, various theories and perspectives have been presented, yet the paradoxes surrounding it only seem to deepen. In *"An Introduction to Time: Universe's Greatest Secret"* I hope to shed light on this enigmatic concept and ignite a deeper curiosity and exploration into its true nature. I look forward to sharing this journey with you and inviting you to question and reflect on your own understanding of time.

@@@@

CHAPTER 2

THE CONCEPT OF TIME

"The only reason for time is so that everything doesn't happen at once.

- Albert Einstein

Time - a concept that has puzzled and intrigued humans since the beginning of our existence. It is something that we all experience, yet it remains a mystery that we are unable to fully comprehend. We measure our lives in terms of time, plan our days and set goals based on the passing of time.

But what exactly is time? Is it a tangible thing that can be measured or is it simply a human construct? Let us examine this never-ending mystery and try to understand the concept of time.

Time has been defined in many ways by diverse cultures and civilizations throughout history. The ancient Greeks believed time to be cyclical, with no beginning or end. They saw time as an endless repetition of events, where everything that has happened will happen again. On the other hand, the concept of linear time emerged in the Judeo-Christian belief system, where time was seen as moving forward in a straight line, with a definite beginning and end.

In modern times, time is defined by scientists as the fourth dimension, along with three dimensions of length, width, and depth. This definition implies that time is something that can be measured and manipulated, just like other physical dimensions. However, despite all these attempts to define time, it remains an enigma that eludes our full understanding.

One of the reasons for this mystery surrounding time is its subjective nature. Time is perceived differently by everyone. Have you ever noticed how an hour can feel like an eternity when you are waiting for something, but it flies by when you are having fun? This is because our perception of time is influenced by our emotions and experiences. *Albert Einstein* once said, "When you sit with a nice girl for two hours, you think it's only a minute. But when you sit on a hot stove for a minute, you think it's two hours." Our perception of time is not constant, making it a difficult concept to grasp.

Another aspect that adds to the complexity of time is its relation to space. The theory of relativity, proposed by *Einstein*, states that time and space are intertwined and cannot be separated. This means that the passage of time is affected by the speed and gravity of an object. For example, time runs slower for objects moving at high speeds and in stronger gravitational fields. This phenomenon has been proven through experiments and has challenged our traditional understanding of time as a linear, uniform concept.

Moreover, the idea of time also varies across different cultures. For some cultures, time is seen as a precious commodity that needs to be used wisely. In contrast, other cultures view time as cyclical, where events repeat themselves in a never-ending cycle. This difference in perception can lead to misunderstandings and conflicts when people from diverse cultures work together. In some cultures, being on time is highly valued, while in others, it is acceptable to be fashionably late.

The concept of time has also been explored extensively in philosophy and religion. The Greek philosopher, *Aristotle*, believed that time was created along with the universe and would cease to exist if the universe were to end. In Hinduism and Buddhism, time is seen as cyclic, with each cycle consisting of creation, preservation, and destruction. Many religions also have specific rituals and practices that revolve around time, such as daily prayers and annual festivals.

In recent years, scientists have made remarkable discoveries about the nature of time that challenge our traditional understanding. The concept of a multiverse, where there are multiple universes coexisting alongside ours, has been proposed by physicists. This concept suggests

AN INTRODUCTION TO TIME

that time may work differently in each universe, further complicating our understanding of this mysterious force.

With all these different perspectives on time, one thing remains certain - it is a limited resource that cannot be regained once lost. Time is constantly moving forward, and every second that passes is a second that we will never get back. This realization can be both terrifying and motivating. It can make us feel the pressure to make the most out of our lives, but it can also make us appreciate the present moment and live in the here and now.

@@@@

CHAPTER 3

THE HISTORY OF TIME

"But ever since the dawn of civilization, people have not been content to see events as unconnected and inexplicable. They have craved an understanding of the underlying order in the world. Today we still yearn to know why we are here and where we come from. Humanity's deepest desire for knowledge is justification enough for our continuing quest. And our goal is nothing less than a complete description of the universe we live in."

— *Stephen Hawking*

Time is a concept that is deeply ingrained in our daily lives. We constantly check the time, plan our schedules accordingly, and even make New Year's resolutions to better manage our time. But have you ever stopped to think about the origin and evolution of this simple yet complex concept?

The history of time is a fascinating and often overlooked subject. It encompasses a multitude of cultures and civilizations, each with their own unique understanding and measurement of time. So, let us take a journey through time and unravel the mysteries of its rich history.

Our journey begins in ancient civilizations such as the Sumerians, Egyptians, and Mayans, who were among the first to develop a structured system for measuring time. The Sumerians, who lived in Mesopotamia around 4000 BC, divided their day into two 12-hour periods, which was later adopted by the Babylonians. These early civilizations used sundials and water clocks to measure the passing of time.

In ancient Egypt, the concept of time took on a more spiritual significance. The Egyptians believed in an afterlife and viewed time as cyclical, with each day representing the journey of the Sun God "*Ra*" through the underworld. They also developed one of the earliest calendars based on the cycles of the Nile River.

AN INTRODUCTION TO TIME

Meanwhile, in Central America, the Mayans developed a highly sophisticated calendar system that combined both solar and lunar cycles. They also believed in cyclical time, with each cycle ending and beginning again in a continuous loop.

The Influence of Astronomy on Timekeeping

As civilizations progressed and developed more advanced technology, their understanding and measurement of time also evolved. One considerable influence on the concept of time was astronomy.

The ancient Greeks were among the first to study astronomy and its relationship to time. They believed that the earth was at the centre of the universe and that the motion of celestial bodies could be used to measure time. This led to the development of the Greek sundial, which had a more accurate measurement of time than previous methods.

Later, in 1582, *Pope Gregory XIII* introduced the Gregorian calendar, which is still widely used today. This calendar was based on the solar cycle and was more accurate in accounting for leap years than its predecessors.

The Invention of Mechanical Clocks

Fast forward to the 14th century, and we see the invention of mechanical clocks in Europe. These clocks were driven by weights and gears and provided a more precise measurement of time than any previous methods. They also revolutionized society, as people could now keep track of time without relying on the position of the sun or stars.

However, it was not until the 17th century when *Galileo Galilei* discovered pendulum motion that clocks became even more accurate. This led to the invention of pendulum clocks, which were used for centuries until they were eventually replaced by quartz clocks in the 20th century.

The Advent of Standard Time

Before the 19th century, each town and city had its own local time based on the position of the sun. This system was known as "local mean time," but it created confusion and made it challenging to coordinate transportation and communication across different regions.

This all changed with the development of railroad networks, which required a standardized method of keeping time. In 1883, representatives from major railway companies came together to establish a standard time system based on Greenwich Mean Time (GMT). This system divided the world into 24 time zones, with each zone representing one hour ahead or behind GMT.

The Measurement of Time Today

Today, we use highly accurate atomic clocks to measure time. These clocks use the frequency of atomic vibrations to keep track of time, making them incredibly accurate – losing only one second every 100 million years. Atomic clocks are used in various applications, from GPS systems to advanced scientific research.

In 1967, the measurement of time was further refined with the introduction of the *International System of Units (SI)*. The SI unit for time is the second, which is defined as the duration of 9,192,631,770 periods of the radiation corresponding to the transition between two energy levels of the cesium-133 atom.

The Concept of Time in Philosophy and Science

Throughout history, philosophers and scientists have attempted to understand and define the concept of time. In ancient Greece, philosophers such as *Plato* and *Aristotle* debated whether time was a continuous flow or a series of discrete moments.

In the 20th century, *Albert Einstein*'s theory of relativity revolutionized our understanding of time. His theory proposed that time is not absolute, but rather varies depending on an observer's frame of reference. This concept is known as "time dilation" and has been proven through experiments and observations in space.

The concept of time also plays a significant role in quantum physics. According to this branch of science, time is not linear but rather a dimension that can bend and curve.

The Future of Time

As we continue to advance technologically and scientifically, our understanding and measurement of time will undoubtedly continue to evolve. Time travel remains a popular topic in science fiction and has sparked debates among scientists about its possibility.

Meanwhile, some scientists are exploring the idea that time may be an illusion, and that past, present, and future all coexist simultaneously. While this concept may seem far-fetched, it challenges us to question our long-held beliefs about time and its perceived linearity.

@@@@

CHAPTER 4

THE PERCEPTION OF TIME THROUGH AGES

"Perception is real even when it is not reality.

- Edward De Bono

Time is a concept that has intrigued humankind since the beginning of civilization. It is an intangible yet ever-present force that dictates the rhythm of our lives. From ancient civilizations to modern societies, the perception of time has evolved and shaped human behaviour and culture. In this chapter, we will take a journey through the ages to explore how the perception of time has changed and influenced human lives.

Ancient Times:

The earliest civilizations, such as the Egyptians, Babylonians, and Greeks, had a cyclical perception of time. They believed that time was a repetitive cycle of birth, growth, decay, and death. This concept was closely tied to their religious beliefs, where time was seen as a manifestation of the divine. The movement of celestial bodies like the sun, moon, and stars were used to measure time, and the calendar was based on natural cycles such as the changing seasons.

For these ancient civilizations, time was not seen as something that could be controlled or managed. It was a force beyond human understanding and control. This perception of time also influenced their attitudes towards life and death. They believed in an afterlife where time would continue in a cyclical pattern.

Middle Ages:

The Middle Ages saw a shift in the perception of time with the rise of Christianity in Europe. Time was now seen as linear, with a beginning and an end. The Christian belief in the creation of the world by God gave rise to the idea of a linear timeline with a predetermined end.

The concept of eternity also gained significance during this period. The belief in an afterlife led people to focus on their actions in the present for the promise of eternal salvation or damnation. Time became a means to an end rather than an endless cycle.

Renaissance:

During the Renaissance period, there was a resurgence of interest in classical Greek and Roman culture. The idea of humanism emerged, where humans were seen as the centre of the universe and capable of shaping their own destiny.

This new perspective gave rise to a more secular approach to time. The focus shifted from the afterlife to the present, and people began to value their time on earth. The invention of the mechanical clock by the Europeans revolutionized the way time was measured and managed. Time was no longer seen as an abstract concept but a measurable and controllable force.

Industrial Revolution:

The Industrial Revolution saw a notable change in the perception of time. With the invention of machines and factories, time became a valuable commodity. The concept of working hours and a strict schedule emerged, and people's lives became increasingly regulated by the clock.

The shift towards a more industrialized society also brought about a change in the perception of time as something that could be quantified and monetized. People were now paid for their time, and the idea of productivity and efficiency became crucial.

20th Century:

The 20th century marked a period of rapid technological advancement that transformed the perception of time once again. With inventions like telephones, radios, and eventually computers, communication and information could be transmitted instantaneously, making the world feel smaller.

The increase in globalization also led to a more standardization of time with the adoption of international time zones. People's lives became busier, and the pressure to manage time efficiently increased.

Present Day:

In today's fast-paced world, the perception of time has become even more complex. The rise of social media and smartphones has made us constantly connected and accessible, blurring the lines between work and personal life.

The idea of "instant gratification" has become prevalent, where people expect things to happen quickly. Time has become a precious resource that is often in short supply, leading to a heightened sense of urgency in everything we do.

Moreover, advancements in technology have also allowed us to manipulate time through various means such as daylight-saving time and time travel in science fiction. These concepts further complicate our understanding of time and its limitations.

The Future:

As we continue to evolve and adapt to a rapidly changing world, it is difficult to predict how our perception of time will continue to evolve. With the rise of artificial intelligence and the possibilities of extending human lifespans, our understanding of time may shift yet again.

@@@@

CHAPTER 5

HUMAN FASCINATION WITH TIME

"Time is a created thing. To say 'I don't have time,' is to say, 'I don't want to.'"

— *Lao Tzu*

Time: the intangible, abstract concept that has captivated the human imagination for centuries. It is a force that cannot be seen or touched, yet it governs our lives in every aspect. We measure our days, our years, our lifetimes through the ticking of a clock and the changing of a calendar. But what is it about time that has captured our fascination and sparked endless debates, theories, and discoveries?

From philosophical musings to scientific breakthroughs, the human fascination with time has shaped our understanding of the world and our place in it. Let us delve deeper into this enduring fascination and explore the wonders and mysteries of time.

The Beginning of Time Measurement

The concept of time measurement can be traced back to ancient civilizations such as the Sumerians, Egyptians, and Babylonians. They used sundials, water clocks, and other primitive tools to track the passage of time. However, it was not until the 14th century that mechanical clocks were invented in Europe, ushering in a new era of measuring time.

With the advancement of technology, more accurate timekeeping devices such as pendulum clocks and atomic clocks were created. Today, we have digital clocks and smartphones that can tell time down to the millisecond.

But why do we feel the need to measure something that is intangible and ever flowing?

The Perception of Time

Time is a fundamental aspect of our daily lives. We use it to plan our activities, keep appointments, and meet deadlines. However, despite its practical importance, our perception of time is highly subjective. Have you ever noticed how time seems to fly by when we are having fun but drags on when we are bored or anxious? This phenomenon is known as "temporal distortion."

Several factors such as emotions, attention, and external stimuli can affect our perception of time. For example, during an adrenaline-pumping activity, our brains are hyper-focused, causing time to appear to speed up. Conversely, in a state of boredom or waiting for something, our minds have less stimuli to process, resulting in time feeling slower.

The Mystery of Time and Its Illusory Nature

One of the most intriguing aspects of time is its elusive and illusory nature. It is constantly moving yet always out of our grasp. As the famous quote by *Benjamin Franklin* goes, "Lost time is never found again." This paradoxical concept has sparked numerous philosophical debates and existential ponderings.

Some philosophers argue that time is an illusion created by our mind to make sense of the world around us. They propose that past, present, and future are mere constructs of human consciousness and that everything exists in an eternal present. This idea challenges our conventional understanding of time as a linear progression.

The human fascination with time also stems from its mysterious nature. Despite centuries of research and technological advancements, we still do not fully understand its essence. The concept of time travel, for example, has captured the imaginations of many, but it remains a theoretical possibility rather than a tangible reality.

The Power of Time

Time holds immense power over us – it can both heal and destroy. It is the ultimate equalizer as it affects everyone equally regardless of age, wealth, or status. Time is also irreversible – once a moment has passed, it can never be relived.

This power of time is evident in the saying "time heals all wounds." As we move further away from a painful event or memory, its emotional intensity decreases with time. Similarly, time is also capable of destroying – a missed opportunity or a lost loved one can never be regained.

The Power of Time in Culture and Art

The human fascination with time has also been reflected in various forms of art and culture throughout history. In literature, authors have explored the concept of time through themes of nostalgia, regret, and the inevitability of aging. In music, time signatures and tempo are used to create a sense of rhythm and flow.

In visual arts, the depiction of time has been a recurring theme. The famous painting "The Persistence of Memory" by *Salvador Dali* is an iconic representation of time with its melting clocks. In cinema, the manipulation of time through techniques such as flashbacks and slow-motion adds depth and complexity to storytelling.

The Limitations of Time

Despite our advancements in measuring and understanding time, it still has its limitations. Time cannot be stopped or reversed, and it constantly moves forward, leaving us with a finite amount of it. This realization has led to the development of various time management techniques and productivity hacks to make the most out of our limited time.

However, the pressure to be constantly productive and make the most out of every moment can also have negative effects on our mental wellbeing. It is important to strike a balance between being mindful of time and allowing ourselves to slow down and appreciate the present.

William Shakespeare said, "Better three hours too soon than a minute too late." The mystery and power of time will continue to intrigue us for generations to come but let us not forget to live in the present while we unravel its wonders.

@@@@

CHAPTER 6

THE FALLACIES OF TIME

"It's one of the great fallacies that time gives much of anything but years and sadness to a man."

- Mike Crittenden

Time is a concept that has been studied and debated for centuries. It is a fundamental aspect of our lives, dictating our daily routines, shaping our memories, and influencing our decisions. However, despite its apparent simplicity, time is a complex and often misunderstood concept. It is subject to various fallacies that can distort our perception and understanding of it. In this chapter, we will explore the fallacies associated with time and delve deeper into their roots and consequences.

Before we begin, it is essential to understand what fallacies are. In simple terms, fallacies are errors in reasoning that can lead to false or invalid conclusions. In the context of time, fallacies refer to misconceptions or flawed beliefs about time that can lead to incorrect understandings or judgments.

The Fallacy of Time as a Constant

One of the most common fallacies associated with time is the belief that it is a constant, unchanging entity. This belief stems from our tendency to measure time in standardized units such as seconds, minutes, and hours. We often perceive time as something that flows at a steady pace, unaffected by external factors.

However, this belief is flawed as time is not constant but relative. The concept of relativity was first introduced by *Albert Einstein* in his theory of special relativity, which states that the laws of physics are the same for all observers in uniform motion. This theory also suggests that time can slow down or speed up depending on the speed and gravity of an object.

For example, imagine two individuals standing on opposite sides of a train track. From one person's perspective, the train is moving at a faster pace than from the other person's viewpoint. This difference in perception is due to the relative motion between the observer and the train.

Moreover, according to *Einstein*'s theory of general relativity, gravity can also affect the passage of time. This phenomenon is known as gravitational time dilation, where time moves slower in areas with stronger gravitational fields. A practical example of this is the time dilation experienced by astronauts in space, where time moves slower due to the weaker gravitational field.

The Fallacy of Time as Linear

Another common fallacy associated with time is the belief that it is a linear concept. This fallacy is rooted in our human experience of time, where we perceive it as a continuous line that moves from the past, through the present, and into the future. We often think of time as a straight line with fixed points, and events occur in a sequential order.

However, this belief does not align with scientific research and theories. The concept of time as linear is an oversimplification of a more complex reality. Quantum physics suggests that time may not be linear but rather a series of interconnected events. This theory challenges our traditional understanding of cause and effect, as events may occur simultaneously or in a non-linear order.

Furthermore, recent studies in psychology have also questioned the linearity of time. Studies have shown that our perception of time can be influenced by our emotions, attention, and memory. When we are engaged in an activity we enjoy, time appears to pass quickly, whereas when we are bored or anxious, it seems to drag on. This phenomenon is known as subjective time dilation and further highlights how our perception of time is not always linear.

The Fallacy of Time as Objective

We often view time as an objective entity that exists independently of our perceptions and experiences. This fallacy stems from our reliance on clocks and calendars to measure and organize our days. We assume that time always progresses at the same pace for everyone, regardless of their circumstances.

However, this belief ignores the fact that our experiences and emotions can significantly impact our perception of time. Our subjective experiences can make time feel faster or slower than it is. For instance, an hour-long lecture may seem like an eternity, while an hour spent with loved ones can fly by in an instant.

Moreover, our cultural and societal norms also influence our understanding of time. In some cultures, time is viewed as a precious resource, and punctuality is highly valued. In contrast, in other cultures, time is perceived as more fluid and flexible, with less emphasis on strict schedules. These cultural differences highlight how time can be subjective and shaped by societal norms and values.

The Fallacy of Time as a Finite Resource

One of the most damaging fallacies associated with time is the belief that it is a limited resource. This fallacy leads to the popular saying, "time is money," where we equate time with productivity and efficiency. We often feel pressure to use our time wisely and make the most of every second.

This belief can lead to feelings of stress, anxiety, and guilt if we feel like we are not using our time effectively. It also creates a sense of urgency to constantly be busy and productive, resulting in burnout and a lack of work-life balance.

However, time is not a finite resource. It is a boundless concept that cannot be controlled or managed. We cannot add more hours to our day or save time for later use. Instead of viewing time as a limited resource, we should focus on managing our priorities and making meaningful use of the time we have.

.@@@@

CHAPTER 7

TIME: LINEAR OR MULTIDIMENSIONAL

"Linear time is a Western invention; time is not linear, it is marvellous tangle where at any moment, points can be selected and solutions invented without beginning or end.

- Lina Bo Bardi

Time is something that we all experience daily. We are constantly aware of its passing, counting down the minutes until an important meeting or eagerly anticipating the arrival of a long-awaited event. But what exactly is time? Is it a linear progression, moving forward in a straight line, or is it more complex and elusive than we can even begin to comprehend?

The concept of time has been a source of fascination and debate for centuries. Philosophers, scientists, and theologians have all grappled with the idea of time and its true nature. And while there may never be a definitive answer, exploring different theories and perspectives can shed light on this mysterious fourth dimension.

Linear Time: The Common Perception

The most common perception of time is that it moves in a linear fashion, from past to present to future. This belief is deeply ingrained in our society and has been shaped by our experiences and interactions with the world around us.

Think about how we measure time – seconds, minutes, hours, days, weeks, years. It all follows a linear progression, with each moment building upon the last. We also have calendars that track the passing of days and seasons, further reinforcing this idea of time as a linear concept.

Additionally, our understanding of cause and effect also supports the notion of linear time. We believe that actions in the past lead to consequences in the present and future. This linear thinking allows us to plan and learn from our mistakes in the past.

Furthermore, our own firsthand experiences also contribute to our belief in linear time. We have memories of events in our past that we can reflect on, and we anticipate future events based on our current experiences.

But is this perception of time as a straight line the only way to look at it?

The Theory of Relativity: A Different Perspective

In the early 20th century, *Albert Einstein* rocked the scientific world with his theory of relativity. According to Einstein, time is relative, meaning that it can be experienced differently by different observers.

One of the key principles of relativity is that time can be influenced by gravity and motion. This theory suggests that time is not a constant and can change depending on the conditions in which it is measured.

For example, a person traveling at high speeds will experience time differently than someone who is stationary. This means that a person on a spaceship traveling at the speed of light would experience time much slower than someone on Earth. This concept has been proven through experiments with atomic clocks on airplanes and satellites, which have shown a slight difference in time compared to those on Earth.

The Spiral Perspective: Time as Cyclical

Another perspective on time comes from ancient cultures and religions, which view time as cyclical rather than linear. In this belief, time is seen as a never-ending cycle, with events repeating themselves in an infinite loop.

The idea of cyclical time can be seen in various cultures' creation stories, in which the world goes through cycles of birth, death, and rebirth. It is also present in Eastern philosophies such as Hinduism and Buddhism, where the concept of reincarnation suggests that we are all trapped in an endless cycle of birth and rebirth.

In this perspective, time is not something that moves forward but rather rotates or spirals around a pivotal point. The past, present, and future are all intertwined and interconnected, with events repeating themselves throughout eternity.

The Illusion of Time: A New Way of Thinking

Some scientists and philosophers propose yet another perspective on time – that it may be an illusion altogether. This idea has gained traction in recent years through the study of quantum physics.

Quantum physicists have discovered that at the subatomic level, particles do not follow traditional laws of cause and effect. Instead, they can exist in multiple states simultaneously, and time becomes a non-linear concept. This challenges our traditional understanding of time and suggests that it may be an illusion created by our minds.

Additionally, some spiritual and metaphysical beliefs suggest that time is an illusion and that all moments are happening simultaneously. This idea is often tied to the concept of the infinite nature of the universe, where everything that has ever happened or will happen is already existing in some form.

But How Does Time Feel to Us?

With all these different theories and perspectives, it can be difficult to understand how time truly functions. And while we may never have a definitive answer, one thing remains constant – our experience of time.

No matter the perspective, we all experience time in a similar way. We perceive it as passing in a linear fashion, with each moment leading to the next. We are also limited by our human perception, which can only comprehend a small fraction of the universe's infinite existence.

In the end, whether time is linear, cyclical, or an illusion, it is a fundamental part of our existence. It shapes our understanding of the world and influences how we live our lives.

So, the true nature of time is something we may never fully comprehend. But that does not stop us from continuing to explore its mysteries and unravelling its secrets. Isn't that what makes the concept of time so intriguing?

@@@@

CHAPTER 8

THE ROLE OF CULTURE AND MEDIA IN UNDERSTANDING OF TIME

"Don't let the fear of the time it will take to accomplish something stand in the way of your doing it. The time will pass anyway; we might just as well put that passing time to the best possible use."

— *Earl Nightingale*

Time is a concept that has baffled and intrigued humans for centuries. It is a fundamental aspect of our lives, yet it is also one of the most elusive concepts to define. From ancient civilizations to modern societies, time has played a crucial role in shaping our cultures and societies. It has been a constant source of fascination for philosophers, scientists, and artists alike. However, one cannot overlook the impact of culture and media on our understanding and perception of time. In this chapter, we will delve deeper into the intricate relationship between culture, media, and the concept of time.

Culture is often described as the shared beliefs, customs, behaviours, and values of a particular group or society. It encompasses everything from our language to our social norms and traditions. Culture is a dynamic entity that evolves over time and is heavily influenced by external factors such as religion, geography, and history. It also plays a significant role in shaping our understanding of time.

In many cultures, time is seen as a cyclical concept rather than a linear one. For example, in ancient Hinduism, time is represented by the concept of 'kalachakra,' which translates to 'wheel of time.' This cyclical view of time is also evident in other cultures such as the Mayan civilization and the indigenous cultures of Australia and Africa. In these societies, time is not seen as something that progresses in a straight line but rather as a continuous cycle of birth, life, death, and rebirth.

On the other hand, Western cultures have a more linear perception of time. Time is seen as finite and constantly moving forward. This concept can be traced back to ancient Greek philosophers such as Aristotle who believed in the concept of 'kairos' – the idea that time is always in motion and cannot be recaptured once it has passed. This linear understanding of time is also reflected in the modern-day concept of time as a limited resource, which has led to the rise of phrases such as 'time is money' and 'time is of the essence.'

Culture not only shapes our perception of time but also influences our relationship with it. For instance, in some cultures, being punctual and adhering to strict timelines is highly valued and seen as a sign of respect. In contrast, other cultures may have a more relaxed attitude towards time and prioritize relationships and experiences over punctuality.

The media also plays a crucial role in shaping our understanding of time. With the advent of technology, media has become an integral part of our daily lives, bombarding us with a constant stream of information and influencing our perception of time. News channels, social media platforms, and other forms of media have made it possible for information to be disseminated at an unprecedented speed, blurring the boundaries between past, present, and future.

One way media impacts our perception of time is through its portrayal of events. In today's fast-paced world, news cycles move quickly, and events that once took days or weeks to unfold are now condensed into a matter of hours. This constant bombardment of information can make it difficult for individuals to process and reflect on events, leading to a sense of time passing by in a blur.

Moreover, media also has the power to manipulate our perception of time by controlling what we see and how we see it. For example, advertisements often use techniques such as fast-paced editing and catchy slogans to create a sense of urgency and scarcity around their products, giving consumers the impression that they must act quickly or risk missing out.

Furthermore, the rise of social media has also had a significant impact on our understanding of time. With platforms such as Facebook and Instagram, individuals are constantly bombarded with images and updates from their friends and peers, creating a sense of *FOMO (fear of missing out)*. This constant need to stay connected and be up to date with what others are doing can create a distorted perception of time, making individuals feel like they are always behind or missing something.

The media also plays a role in shaping cultural perceptions of time. For instance, the American culture's emphasis on productivity and efficiency can be attributed in part to the portrayal of successful individuals and businesses in the media. This constant exposure to the 'hustle culture' has led to a societal pressure to always be busy and productive, leading to a sense of time being a limited resource that must be utilized wisely.

In addition to shaping our understanding of time, media also has the power to influence our sense of time. Music, for example, has been known to have a profound effect on our perception of time. A fast-paced song can make time seem to fly by, while a slow and soothing tune can make it feel like time has slowed down.

@@@@

CHAPTER 9

PHILOSOPHY OF TIME

"It doubtless seems highly paradoxical to assert that Time is unreal, and that all statements which involve its reality are erroneous. Such an assertion involves a far greater departure from the natural position of mankind than is involved in the assertion of the unreality of Space or of the unreality of Matter. So decisive a breach with that natural position is not to be lightly accepted. And yet in all ages the belief in the unreality of time has proved singularly attractive."

— *John McTaggart*

Time has always been a fundamental aspect of human existence. It is something that we constantly experience and yet struggle to fully understand. From the ticking of a clock to the passing of days, time is an integral part of our daily lives. However, beyond its practical implications, time is also a complex and intriguing concept that has sparked numerous philosophical debates throughout history.

In this chapter, we will delve into the philosophy of time, exploring its various dimensions and shedding light on some of the most thought-provoking theories surrounding it.

The Perception of Time

One of the most fascinating aspects of time is its subjective nature. We all experience time differently – sometimes it seems to fly by, while at other times it may seem to drag on forever. This subjectivity has led philosophers to question whether time is an objective reality, or a mere illusion created by our minds.

The Greek philosopher, *Aristotle,* argued that time is a measure of change, and thus it exists independently from our perception of it. According to him, time is a continuous and infinite succession of moments that can be measured through motion and change. On the other hand, the philosopher *Immanuel Kant* proposed that time is not an external reality but rather a 'pure intuition' that is imposed by our minds to make sense of our experiences.

The concept of 'presentism' also plays a crucial role in understanding the perception of time. Presentism states that only the present moment exists, and the past and future are merely mental constructs. This idea challenges our common belief that time flows in a linear manner from past to present to future. Instead, presentism suggests that all moments exist simultaneously, and it is our consciousness that creates a sense of progression.

The Nature of Time

Apart from its subjective nature, philosophers have long debated about the ontological status of time – whether it is real or simply an illusion. Some argue that time is a fundamental reality that exists independently from human perception, while others believe it is a human construct.

The German philosopher, *Gottfried Wilhelm Leibniz*, proposed the concept of 'relational time' which states that time is a fundamental aspect of the universe, and it arises from the relationships between objects. This means that time does not exist on its own but rather as a product of the interactions between objects.

On the other hand, the theory of 'eternalism' suggests that time is a dimension that encompasses all moments – past, present, and future – as equally real. Eternalism states that all moments exist simultaneously in a 'block universe', and our perception of time as a linear progression is an illusion.

The Illusion of Time

The concept of an illusionary nature of time has been a subject of great interest among philosophers. Many have argued that our perception of time as a linear progression is nothing more than an illusion created by our minds.

The philosopher *J.M.E McTaggart* proposed the theory of 'The Unreality of Time' which suggests that past, present, and future do not actually exist. According to him, time is not an objective reality but rather a subjective experience created by our consciousness. He believed in the idea of an 'ever-present now', where all moments exist simultaneously and there is no real passage of time.

Furthermore, the theory of 'time dilation' in physics has also added to the debate about the illusionary nature of time. This theory states that time moves at different speeds for different observers depending on their relative motion and gravitational pull. This means that the concept of time itself is subjective and can vary from one observer to another.

The Philosophy of Time and Ethics

The concept of time has also been linked to ethical considerations. The idea that our time on earth is limited has led philosophers to ponder the significance of utilizing this finite resource effectively.

The philosopher *Martin Heidegger* believed that human beings often get lost in the everyday business of life and fail to appreciate the true nature of time. He argued that we should embrace the finitude of human existence and live in the present moment, rather than being consumed by thoughts of the past or future.

Similarly, the philosopher *Seneca* believed in making the most of our time by living a fulfilling and virtuous life. He emphasized the importance of self-reflection and constantly striving to better oneself, for time is a precious resource that should not be wasted.

@@@@

CHAPTER 10

SCIENCE BEHIND TIME

"If time travel is possible, where are the tourists from the future?"

- Stephen Hawking

Time is a concept that governs our daily lives. We use it to schedule our commitments, plan our activities, and keep track of our daily routines. From the moment we wake up in the morning to the time we go to bed at night, time plays a crucial role in how we live our lives. But have you ever stopped to think about the science behind the measurement of time? How did humans come up with the idea of dividing time into seconds, minutes, and hours? In this chapter, we will explore the fascinating science behind the measurement of time.

The Early History of Timekeeping

The concept of measuring time can be traced back to ancient civilizations. The earliest form of timekeeping was based on the observation of natural phenomena such as the movement of the sun, moon, and stars. For instance, the ancient Egyptians used obelisks to track the movement of the sun and used sundials to divide the day into 12 hours. The ancient Greeks also used sundials, but they were the first to divide the day into 24 equal hours.

The Birth of Clocks

The invention of mechanical clocks in the 14th century revolutionized the way we measure time. These clocks were based on the pendulum mechanism, which allowed for more accurate and consistent timekeeping. However, it was not until the 17th century that *Galileo Galilei* discovered that a pendulum's swing is not entirely consistent due to factors such as air resistance and friction. This discovery led to the invention of the escapement mechanism by Christian Huygens, which further improved the accuracy of clocks.

The Advent of Atomic Clocks

While mechanical clocks were a significant advancement in timekeeping, they were still not entirely accurate. It was not until 1955 that scientists developed atomic clocks, which are currently considered the most accurate way to measure time. Atomic clocks work by measuring the vibrations of atoms, specifically the *Cesium (Caesium)* atom, which oscillates at a frequency of 9,192,631,770 cycles per second. This incredible accuracy has allowed scientists to measure time to an accuracy of one second in 100 million years.

The Definition of a Second

In 1967, the International System of Units (SI) defined the second as the duration of 9,192,631,770 periods of the radiation corresponding to the transition between two hyperfine levels of the ground state of the cesium-133 atom. This definition has remained unchanged since then and is used as the standard for defining one second. This means that all other units of time, such as minutes and hours, are based on this standard definition of a second.

The Role of Relativity in Timekeeping

In addition to atomic clocks, *Albert Einstein's* theory of relativity has also played a significant role in our understanding and measurement of time. According to *Einstein's* theory, time is not absolute and can vary depending on the observer's frame of reference. This means that time can pass at different rates for two observers depending on their relative velocity or proximity to a massive object such as a planet or star. The effects of relativity are not noticeable in our daily lives, but they are crucial for accurate timekeeping in fields such as GPS navigation and satellite communication.

The Leap Second

Despite the incredible accuracy of atomic clocks, they still need to be adjusted periodically to account for the slight variations in Earth's rotation. This is done by adding a leap second to Coordinated Universal Time (UTC), which is based on atomic time but adjusted to match solar time. A leap second is added every 18 months on either December 31st or June 30th.

The Future of Timekeeping

With advancements in technology and our understanding of time, scientists continue to push the boundaries of accurate timekeeping. Currently, the most accurate atomic clock, the National Institute of Standards and Technology-F1, is expected to neither gain nor lose a second for approximately 50 million years. However, researchers are working on developing even more precise clocks, such as optical atomic clocks, which use lasers to measure the vibrations of atoms and have the potential to be accurate to a second in 100 billion years.

The Science Behind Our Perception of Time

While the measurement of time has been thoroughly studied and defined by science, our perception of time is a much more complex and subjective concept. Have you ever noticed that time seems to pass by faster when we are having fun or slower when we are bored? This is because our brains process time differently depending on our emotional state and level of engagement. Studies have shown that activities that require more attention and engagement tend to make time feel longer, while routine tasks can make time seem to fly by.

@@@@

CHAPTER 11

UNDERSTANDING OF TIME FROM ARISTOTLE TO EINSTEIN AND BEYOND

"Einstein never accepted that the universe was governed by chance; his feelings were summed up in his famous statement God does not play dice."

- Stephen Hawking

Time is a fundamental aspect of our existence and has been a subject of fascination and contemplation for centuries. From ancient philosophers to modern-day scientists, the concept of time has been studied and debated, leading to various theories and understandings. In this chapter, we will discuss the understanding of time from *Aristotle* to *Einstein* and how it has shaped our perception of the world.

Aristotle, the Greek philosopher, was one of the first to develop a comprehensive theory of time. He believed that time was a continuous and uniform flow, with past, present, and future coexisting simultaneously. According to *Aristotle*, time was a measurement of change and motion in the physical world. He argued that time cannot be divided into smaller units as it is indivisible and continuous.

Aristotle's theory of time had a considerable influence on medieval thinking, where time was considered a linear progression from the past to the present. This idea was rooted in the notion of divine providence, where God had predetermined the course of events in time. This concept of time as a linear progression continued until modern times.

However, with the rise of science, there came a shift in the understanding of time. In the 17th century, *Isaac Newton* proposed the concept of absolute time. He argued that time was an independent entity that exists, regardless of any external factors or events. This theory had a profound impact on physics and laid the foundation for our modern understanding of time.

But it was not until the early 20th century when *Albert Einstein*'s theory of relativity challenged *Newton*'s concept of absolute time. According to Einstein, time is not an independent entity but is relative to an observer's frame of reference. This means that time can be perceived differently by different observers depending on their relative speeds or positions.

Einstein's theory also introduced the concept of spacetime, where space and time are interconnected and cannot be separated. This revolutionary idea changed our understanding of time, as it showed that it is not just a linear progression but a dynamic and flexible entity. It also opened new possibilities for understanding the universe and its workings.

One of the most significant implications of *Einstein*'s theory of relativity was the concept of time dilation. This phenomenon states that time passes slower for objects moving at high speeds compared to those at rest. This has been proven through experiments with atomic clocks on spaceships, showing that time has indeed slowed down for objects in motion.

Another important aspect of *Einstein*'s theory is the notion of the speed of light being constant in all frames of reference. This means that the speed of light is the same regardless of an observer's relative speed or position. This concept led to the idea of time being relative to space, and both are interconnected.

Einstein's theory also had a significant impact on our perception of time on a cosmic scale. According to his theory, gravity can warp space and time, causing them to bend and curve. This idea has been confirmed through observations of gravitational lensing, where the light from distant objects is bent due to the curvature of spacetime by massive objects.

Furthermore, through his famous equation $E=mc^2$, *Einstein* showed that mass and energy are interchangeable, leading to the understanding that time can also be influenced by energy. This has been demonstrated through experiments with atomic clocks in high-energy environments, where time has been shown to speed up or slow down depending on the amount of energy present.

Einstein's theory of relativity revolutionized our understanding of time and paved the way for modern physics. It also challenged traditional notions of linear progression and absolute time, showing that time is a complex and dynamic entity.

@@@@

CHAPTER 12

SCIENTIFIC THEORIES TO EXPLAIN TIME

"The eventual goal of science is to provide a single theory that describes the whole."

- Stephen Hawking

Time is a concept that has captivated human minds for centuries. It is a fundamental aspect of our existence, dictating our daily routines and permeating every aspect of our lives. But despite its ubiquitous presence, time remains a mysterious and elusive concept. How do we define it? Is it a physical entity or a human construct? These are just some of the questions that have puzzled philosophers, scientists, and ordinary individuals alike.

Throughout history, various scientific theories have been proposed to explain the concept of time. These theories have evolved over time as our understanding of the universe has deepened. In this chapter, we will examine the scientific theories of time and explore their implications for our understanding of the universe.

The Classical Theory of Time

The classical theory of time, also known as Newtonian time, was first proposed by *Sir Isaac Newton* in the 17th century. According to this theory, time is a fundamental and absolute entity that exists independently of any external factors. In other words, time is constant and unchanging, and it flows at a constant rate.

This theory was based on *Newton*'s laws of motion and gravitation, which stated that time is an essential component in understanding the physical world. *Newton* believed that time was an objective reality that could be measured with a clock. He also introduced the concept of absolute space, which was believed to be a fixed and unchanging framework in which events occur.

The classical theory of time had a profound impact on our understanding of the universe and laid the foundation for modern physics. However, this theory was later challenged by *Albert Einstein*'s theory of relativity.

The Theory of Relativity

Einstein's theory of relativity revolutionized our understanding of time and space. This theory proposed that time is not an absolute entity but rather a relative one that depends on an observer's frame of reference. In other words, time is not constant, and it can be affected by factors such as gravity and acceleration.

Einstein's famous equation, $E=mc^2$, showed the relationship between mass and energy and proposed that time and space are intertwined. According to this theory, massive objects create a curvature in space-time, causing time to slow down in their presence. This phenomenon is known as time dilation and has been proven through experiments with atomic clocks on airplanes and satellites.

The theory of relativity also introduced the concept of spacetime, where space and time are no longer viewed as separate entities but rather as interconnected dimensions. This theory challenged the classical view of absolute space and time and paved the way for a new understanding of the universe.

Quantum Theory of Time

The concept of time takes on a whole new meaning in the world of quantum mechanics. According to this theory, the behaviour of particles at the subatomic level is described as a wave function, which allows particles to exist in multiple states simultaneously.

In quantum mechanics, time is a directionless phenomenon, and particles are viewed as existing in a state of superposition until they are observed or measured. This means that the past, present, and future all exist simultaneously in the quantum world.

The concept of time in quantum mechanics has sparked many debates and discussions among scientists. Some argue that it challenges our understanding of causality, while others see it as evidence of a multiverse where every outcome exists simultaneously.

The Arrow of Time

One of the most intriguing questions about time is its directionality. Why do we perceive time as moving forward? This question is known as the arrow of time, and it continues to puzzle scientists.

According to the second law of thermodynamics, the entropy or disorder in a closed system will always increase over time. This means that our universe is constantly moving towards a state of maximum disorder or equilibrium. This phenomenon is known as the arrow of time, and it provides a direction for the flow of time.

However, this raises the question of why the universe started in a state of low entropy and has been moving towards higher entropy ever since. Some scientists believe that this is due to the *Big Bang*, where the universe began in a state of high order and has been expanding and increasing in entropy ever since.

The concept of time and its directionality continue to be a topic of ongoing research and debate among scientists.

Implications for our Understanding of the Universe

The scientific theories of time have not only changed our understanding of time itself but have also had a significant impact on our understanding of the universe. The concept of spacetime in relativity has challenged our view of a fixed and absolute universe and has shown that space and time are interconnected dimensions that can be affected by massive objects.

The quantum theory of time has opened new possibilities for our understanding of reality, challenging our traditional views of cause and effect. It has also raised questions about the existence of parallel universes and the nature of time itself.

Furthermore, these theories have implications for how we perceive the past, present, and future. In classical physics, these are viewed as distinct and separate entities, while in quantum mechanics, they exist simultaneously. This challenges our intuitive understanding of time and its directionality.

@@@@

CHAPTER 13

EXPERIMENTS TO COMPREHEND TIME

"Time moves in one direction, memory in another."

- William Gibson

Time, a concept that has puzzled and intrigued scientists and philosophers for centuries. What exactly is time? Is it a physical entity or just a human construct? These are questions that have led to numerous experiments to understand and explain the concept of time. In this chapter, we will explore some of the most significant experiments conducted to unravel the mystery of time.

The First Attempts at Studying Time

The earliest recorded experiments on time can be traced back to ancient civilizations such as the Egyptians, Greeks, and Babylonians. They used sundials and water clocks to measure time based on the movement of celestial bodies. However, it was not until the 17th century that scientists began to delve deeper into the concept of time.

Galileo Galilei, a pioneer in the field of modern physics, conducted a thought experiment known as the "Galilean Relativity." In this experiment, he proposed that time is relative and can be affected by motion. He argued that time is not a fixed universal constant but rather depends on the observer's frame of reference. This idea laid the foundation for future experiments on time.

The Experiment That Changed Everything: The Speed of Light

In the late 19th century, *Albert Michelson* **and** *Edward Morley* conducted an experiment to measure the speed of light. The results of this experiment proved to be a turning point in our understanding of time.

According to *Einstein*'s theory of special relativity, the speed of light is constant in a vacuum and is unaffected by an observer's frame of reference. This means that no matter how fast an observer is moving, they will always measure the speed of light as 299,792,458 meters per second (m/s). This concept challenged *Newton*'s laws of motion and sparked a new era in physics.

The concept of time dilation emerged from this experiment, which suggests that an object's time is affected by its velocity. The faster an object moves, the slower its time appears to an outside observer. This phenomenon has been proven through numerous experiments, including the famous "twin paradox," where one twin travels at high speeds in space while the other stays on Earth. When the space-traveling twin returns, they would have aged significantly less than their Earth-bound twin.

The famous equation $E=mc^2$, derived by *Einstein* based on his theory of special relativity, also explains how time is relative to an object's energy and mass. It suggests that as an object's speed approaches the speed of light, its mass increases, and time slows down.

The Clock Paradox and Time Travel

Another experiment that further explored the concept of time dilation was the "clock paradox." In this experiment, a clock is taken on a journey at high speeds, and when it returns, it shows less elapsed time than clocks on Earth.

This paradox has led to the development of theories around time travel. According to the theory of general relativity, if an object travels near the speed of light or in a strong gravitational field, it can bend space-time and create a "time loop." This means that time travel could be possible in theory, but it would require immense amounts of energy and technology that is currently beyond our capabilities.

Quantum Entanglement and Time

Quantum entanglement is a phenomenon where two particles become connected in such a way that any change in one particle affects the other, no matter how far apart they are. This suggests that there is a connection between these particles that transcends space and time.

In 2013, physicists from the University of Vienna conducted an experiment that showed quantum entanglement can occur across time. They were able to entangle two photons separated by a significant distance and measure their entanglement over a period of 10 microseconds. This experiment suggests that time may not be as linear as we perceive it to be.

The Arrow of Time

One of the most intriguing aspects of time is its direction. Why do we experience time moving forward and not backward? This question has led to the development of the "arrow of time" theory.

In 1927, physicist *Arthur Eddington* proposed the idea that the arrow of time is determined by the increase of entropy, a measure of disorder in a system. This means that over time, everything in the universe becomes more disordered, and this is the reason we perceive time as moving forward.

Recent experiments have attempted to challenge this theory, such as a study conducted by researchers at the University of California, Berkeley. They showed that at a quantum level, particles can move backward in time and reorganize themselves to decrease their entropy. While this experiment does not disprove the arrow of time theory, it does raise questions about its validity.

The Grandfather Paradox and the Nature of Time

One of the most famous thought experiments related to time is known as the "Grandfather Paradox." It questions whether it would be possible to go back in time and change an event that has already occurred.

This paradox raises questions about the nature of time itself. Some theories suggest that all events in time have already occurred simultaneously, and our perception of time is just an illusion. Others argue that there are multiple timelines or parallel universes in which every outcome exists.

While we may never have a definite answer to these questions, experiments such as "delayed-choice quantum eraser" have shown that particles can behave differently depending on whether their behaviour will be observed in the future. This suggests that future events could potentially affect present ones and adds to the complexity of understanding time.

@@@@

CHAPTER 14

UNDERSTANDING THE THEORY OF RELATIVITY

"Time is an illusion."

— *Albert Einstein*

The theory of Relativity is one of the most revolutionary and groundbreaking theories in the history of modern physics. It was developed by *Albert Einstein* in the early 20th century and completely changed the way we understand space, time, and gravity. This theory has not only had a significant impact on scientific advancements but also on our daily lives.

So, what is the theory of Relativity and why is it so influential? Let us delve deeper and try to understand this complex and fascinating theory.

What is the theory of Relativity?

The theory of Relativity has two main components: the Special Theory of Relativity and the General Theory of Relativity. The Special Theory of Relativity, published in 1905, explains the relationship between space and time for objects moving at a constant speed in a straight line. It also introduced the famous equation $E=mc^2$, which states that energy (E) is equal to mass (m) multiplied by the speed of light (c) squared.

The General Theory of Relativity, published in 1915, builds upon the Special Theory of Relativity, and includes gravity as a fundamental force. It explains that gravity is not a force between masses, as *Newton*'s theory suggests, but rather a curvature of space and time caused by massive objects.

Together, these two theories completely revolutionized our understanding of the universe and paved the way for new developments in physics.

Key concepts of the theory of Relativity

1. Space and Time are Relative

According to the Special Theory of Relativity, space and time are not absolute but rather relative to the observer's frame of reference. This means that measurements of space and time can vary depending on how an observer is moving relative to an event.

To better understand this concept, imagine two people standing on a train platform. One person observes a train passing by at a constant speed while the other person is on the train. The person on the platform would measure a longer distance and a shorter time for the train to pass, while the person on the train would measure a shorter distance and a longer time. This is due to the relative motion between the observer and the event.

2. The Speed of Light is Constant

Einstein's theory also states that the speed of light is constant, regardless of an observer's frame of reference. This means that no matter how fast an observer is moving, they will always measure the speed of light to be 299,792,458 meters per second.

This concept challenged the existing Newtonian theory, which suggested that the speed of light was relative to an observer's frame of reference. Einstein's theory was later proven correct through experiments and has become a fundamental principle in modern physics.

3. Mass and Energy are Equivalent

The famous equation $E=mc^2$ introduced in the Special Theory of Relativity shows that mass and energy are equivalent and interchangeable. This means that mass can be converted into energy and vice versa.

This concept has been proven through nuclear reactions, where a small amount of mass is converted into a tremendous amount of energy. It has also played a crucial role in understanding the behaviour of particles in atomic and subatomic levels.

4. Gravity is a Curvature of Space-Time

The General Theory of Relativity explains gravity as a curvature of space-time caused by massive objects. In simpler terms, this means that objects with a larger mass have a stronger gravitational pull and can bend space and time around them.

This idea was revolutionary as it challenged *Newton*'s theory that gravity was a force acting between masses. The General Theory of Relativity provided a new understanding of gravity and its effects on objects in space.

Implications and Applications of the theory of Relativity

The theory of Relativity has had a profound impact on our understanding of the universe and has led to many advances in science and technology. Some of its implications and applications include:

1. GPS Technology

The Global Positioning System (GPS) is a technology that uses satellites to determine the location of a receiver on Earth. GPS relies on precise measurements of time and location, which would not be possible without the theory of Relativity. The satellites used in GPS must consider the effects of both Special and General Relativity in their calculations to provide accurate results.

2. Gravitational Waves

Einstein's theory predicted the existence of gravitational waves, which are ripples in the fabric of space-time caused by massive objects moving at high speeds. In 2015, scientists were able to detect these waves using advanced equipment, further confirming the validity of the theory of Relativity.

3. Nuclear Energy

The equation E=mc² has played a crucial role in the development of nuclear energy. This equation explains how a small amount of mass can be converted into a significant amount of energy, as seen in nuclear reactions.

4. Black Holes

The theory of Relativity has also helped us understand and study black holes, which are regions in space where gravity is so strong that not even light can escape. These objects are predicted by *Einstein*'s theory and have been observed through advanced telescopes.

@@@@

CHAPTER 15

QUANTUM MECHANICS

"Life is strong and fragile. It is a paradox... It is both things, like quantum physics: It is a particle and a wave at the same time. It all exists all together."

- Joan Jett

Quantum mechanics has been a topic of fascination and intrigue for scientists and non-scientists alike since its inception in the early 20th century. This revolutionary theory has completely changed our understanding of the physical world, challenged traditional concepts, and shed light on the fundamental building blocks of our universe. One of the most overwhelming aspects of quantum mechanics is its relationship with time, as it poses a complex and often perplexing puzzle for scientists to unravel. In this chapter, we will delve into the intricacies of quantum mechanics related to time and explore the mysteries that lie within.

To understand the connection between quantum mechanics and time, we must first understand the concept of time itself. Time is a fundamental aspect of our daily lives and is commonly defined as a measure of duration, progression, and change. However, in the realm of quantum mechanics, time holds a different meaning. According to this theory, time is not a continuous flow but rather a discrete quantity that is affected by the laws of quantum mechanics.

One of the most significant contributions of quantum mechanics to our understanding of time is the concept of superposition. This refers to the ability of particles to exist in multiple states simultaneously until they are observed or measured. This means that at a subatomic level, particles can exist in two or more places at once, making it impossible to pinpoint their exact location or state at any given moment. This idea challenges our traditional understanding of time as a continuous progression, as superposition suggests that particles can exist in multiple states at different points in time.

Another key concept in quantum mechanics related to time is the Uncertainty Principle, proposed by *Werner Heisenberg* in 1927. This principle states that it is impossible to know with certainty both the position and momentum of a particle at the same time. This means that there will always be uncertainty in our measurements, leading to an uncertain understanding of the passage of time. This concept also challenges the traditional idea of cause and effect, as it suggests that the outcome of an event cannot be completely determined based on its initial conditions.

One of the most puzzling aspects of quantum mechanics related to time is the phenomenon of quantum entanglement. This refers to the strong correlation between two or more particles, where the state of one particle is directly affected by the state of another, even if they are separated by vast distances. This connection between particles appears to defy the laws of classical physics and reinforces the idea that time is not a linear progression in the quantum world. In fact, some scientists have proposed that entanglement may provide a clue to understanding the true nature of time itself.

The renowned physicist, *Albert Einstein*, famously referred to quantum entanglement as "spooky action at a distance." This term captures the strange and perplexing nature of this phenomenon, which has been proven to exist through many experiments. The concept of entanglement raises questions about how particles communicate with each other and what this means for our understanding of time and space.

The relationship between quantum mechanics and time also reveals itself through the concept of time dilation. This refers to the slowing down or speeding up of time relative to an observer's perspective. According to *Einstein*'s theory of relativity, this phenomenon occurs due to differences in velocity and gravity. In the quantum world, time dilation can occur due to the strong gravitational pull of massive objects, such as black holes. This can result in extreme distortions of time, making it almost impossible for us to understand.

Furthermore, quantum mechanics also challenges our understanding of time through the concept of parallel universes. According to some interpretations of quantum mechanics, every decision or outcome creates a new universe branching off from the current one. This means that there could be infinite parallel universes existing at any given moment, each with its own perception of time. This idea adds another layer of complexity to our understanding of time and the role it plays in the quantum realm.

@@@@

CHAPTER 16

THE ARROW OF TIME

"The arrow of time does not move forward forever. There is a phase in the history of the universe where you go from low entropy to high entropy. But then, once you reach the locally maximum entropy you can get to, there is no more arrow of time."

- Sean M. Carroll

Time is a concept that has fascinated humans for centuries. We measure it, we try to control it, yet it remains a mystery that continues to elude us. One of the most intriguing aspects of time is the concept of the "Arrow of Time", which has captured the attention of philosophers, scientists, and even artists. In this chapter, we will take a deep dive into the elusive nature of time and explore the intricacies of the Arrow of Time.

What is the Arrow of Time?

The term "Arrow of Time" was first coined by British astronomer *Arthur Eddington* in 1927. It refers to the one-way direction in which time moves, from the past to the present and into the future. This concept is based on the idea that time has a definite direction and can only move forward, never backward. The Arrow of Time is associated with the second law of thermodynamics, which states that in any isolated system, entropy (or disorder) will always tend to increase.

To understand this concept better, let us imagine a simple scenario – you are watching a video of a glass shattering into pieces. If you were to reverse the video, you would see the glass miraculously coming back together, defying all laws of physics. This is because in our daily experience, we only observe events moving forward in time. The shattered glass represents a state of higher entropy, while the intact glass represents a state of lower entropy. As per the second law of thermodynamics, it is highly unlikely for a system to spontaneously move from a state of higher entropy to a state of lower entropy.

The Arrow of Time in Physics

In physics, time is a fundamental dimension that is intertwined with space. The theory of relativity postulates that time and space are not absolute but are relative concepts that depend on an observer's frame of reference. This means that the measurement of time can vary for different observers moving at different speeds.

The concept of the Arrow of Time is also closely related to the idea of causality, which states that every event has a cause and effect. The Arrow of Time dictates that causality can only flow in one direction – from the past to the present and into the future.

Another important aspect of the Arrow of Time is its relationship with the expansion of the universe. In 1929, *Edwin Hubble* discovered that the universe is expanding, and this discovery led to the development of the Big Bang theory. According to this theory, the universe began as a singularity and has been expanding ever since. The Arrow of Time aligns with this expansion, with events moving from a state of lower entropy (the singularity) to higher entropy (the expanding universe).

The Mystery and Controversy surrounding the Arrow of Time

Despite its prominent role in physics, the concept of the Arrow of Time remains a mystery. One of the biggest controversies surrounding it is the possibility of time flowing backward or "time-reversed causality." This would mean that events in the future could have an impact on events in the past, contradicting our understanding of causality and the Arrow of Time. Some scientists have proposed theories such as "retro causality" or "closed time like curves" to explain this phenomenon, but these ideas are still highly debated and not widely accepted in the scientific community.

Another intriguing aspect of time is its perception by living beings. Humans experience time in a linear fashion, with a clear distinction between past, present, and future. However, studies have shown that animals perceive time differently. For example, some animals can sense impending disasters before they happen, suggesting that they may have a different perception of time compared to humans.

The Arrow of Time in Philosophy and Art

The Arrow of Time has also been a subject of philosophical discussions throughout history. Philosophers have debated whether time is an illusion or a fundamental aspect of reality. Some argue that the concept of time is a human construct and may not exist outside of our perception. Others believe that time is a fundamental aspect of the universe and is interconnected with all other dimensions.

Artists have also been fascinated by the concept of the Arrow of Time and have incorporated it into their works. For example, the famous painting "Persistence of Memory" by *Salvador Dali* depicts melting clocks, representing the fluidity and subjective nature of time. Similarly, many writers have explored the idea of time travel, where the Arrow of Time can be manipulated, leading to complex and thought-provoking storylines.

The Role of Technology in Understanding the Arrow of Time

Advancements in technology have allowed us to delve deeper into the mysteries of time. For example, the development of atomic clocks has enabled us to measure time more accurately than ever before. The study of black holes and gravitational waves has also given us insights into how time behaves in extreme conditions.

Moreover, scientists are currently conducting experiments to test the theory of relativity and its implications on time. For instance, the Large Hadron Collider in Switzerland is being used to study particles moving at high speeds, providing us with valuable information about the behaviour of time at different velocities.

@@@@

CHAPTER 17

TIME PERCEPTION

"Time is basically an illusion created by the mind to aid in our sense of temporal presence in the vast ocean of space. Without the neurons to create a virtual perception of the past and the future based on all our experiences, there is no actual existence of the past and the future. All that there is, is the present."

— *Abhijit Naskar*

Time is a concept that has captivated humans since ancient times. It is a fundamental aspect of our daily lives, yet it is true nature and perception remain a mystery. We are all familiar with the saying "time flies when you're having fun," but have you ever wondered why some moments seem to last forever while others pass by in the blink of an eye? Let us delve into the fascinating world of time perception and try to unravel its complexities.

What is Time Perception?

Time perception can be defined as the subjective experience of the passage of time. It is our individual interpretation of how fast or slow time seems to be moving. This can vary from person to person and even from moment to moment. Time perception is influenced by a variety of factors, including our emotions, attention, and memory.

The Illusion of Time

One of the most intriguing aspects of time perception is its illusionary nature. We often feel like we have a firm grasp on the passing of time, but upon closer examination, we realize that our perception can be easily manipulated. For example, have you ever noticed that time seems to slow down when we are anticipating something or hurry up when we are dreading it? This phenomenon is known as the "oddball effect" and has been studied extensively by psychologists.

Our internal clock also plays a significant role in the illusion of time. It is responsible for regulating our daily rhythms, such as sleep patterns and hunger cues. Our internal clock can be influenced by external factors such as light and temperature, which can alter our perception of time. This explains why we tend to lose track of time when engrossed in an activity or when changing time zones while traveling.

Emotions and Time Perception

Our emotions can have a powerful impact on our perception of time. When we are experiencing strong emotions, such as fear or excitement, time tends to slow down. Studies have shown that individuals who are in life-threatening situations often report that time seemed to move in slow motion. This is because our brain is working overtime, processing information, and trying to keep us safe, which results in a heightened sense of time.

On the other hand, when we are bored or engaged in a monotonous activity, time seems to speed up. This can be attributed to the lack of stimulation and our brain's tendency to switch to autopilot mode. Have you ever noticed how quickly time passes while binge-watching your favourite TV show, but how slowly it seems to move during a tedious meeting?

The Role of Attention and Memory

Our ability to pay attention to the present moment can also influence our perception of time. When we are fully engaged and focused on an activity, time seems to fly by. This is because our brain is not actively processing the passage of time, and we are fully immersed in the present moment. On the other hand, when our minds wander and we are not fully engaged, we tend to experience a slower perception of time.

Our memory also plays a crucial role in shaping our perception of time. When we look back on a past event, our memory can distort our perception of how long it lasted. We often remember significant moments in our lives as lasting longer than they did. This is because our brain tends to focus on the most memorable parts of an event, rather than the entire duration.

Cultural Influences

It is worth mentioning that cultural influences can also have an impact on our perception of time. In some cultures, time is seen as a valuable commodity that must be used efficiently. In contrast, in others, time is viewed as fluid and less rigidly structured. These cultural differences can influence how individuals perceive and prioritize their time.

The Impact of Technology

In today's fast-paced world, technology has undoubtedly had a significant impact on our perception of time. With the constant access to information and instant gratification, our attention spans have shortened, and we expect things to happen quickly. This mindset can lead to a distorted perception of time, where we feel like we never have enough of it. Additionally, technology has also blurred the lines between work and leisure time, making it harder for us to disconnect and truly relax.

The Importance of Mindfulness

In a world where time is constantly slipping away, practicing mindfulness can help us slow down and appreciate the present moment. By being fully present and engaged in our daily activities, we can improve our perception of time and reduce stress and anxiety. Mindfulness also allows us to disconnect from distractions and be more intentional with our time.

As author *William Penn* once said, "Time is what we want most but what we use worst."

@@@@

CHAPTER 18

PSYCHOLOGICAL ASPECTS OF PERCEPTION OF TIME

"They always say time changes things, but you actually have to change them yourself."

— *Andy Warhol*

Time is a concept that has puzzled us for centuries. It is both a fundamental aspect of our daily lives and a mysterious phenomenon that we struggle to fully comprehend. We try to measure it, control it, and even manipulate it, but it seems to slip away from our grasp. However, recent research in psychology and neuroscience has shed light on the complex relationship between our minds and time perception. In this chapter, we will explore the fascinating connection between psychology and time perception, and how various psychological and neurological factors influence our understanding of time.

Our perception of time is subjective and can vary from person to person. Have you ever noticed how time seems to pass by quickly when you are having fun, but drags on when you are bored? This phenomenon is known as "time dilation," where our perception of time is distorted by our emotions and experiences. Our sense of time is not an accurate reflection of the clock or calendar, but rather a product of our minds.

One of the key psychological factors that affect our understanding of time is attention. Our attention plays a crucial role in how we process and perceive time. When we are fully engaged in an activity, like reading a book or playing a video game, we tend to lose track of time because our attention is fully absorbed in the present moment. On the other hand, when we are bored or have nothing to do, our minds tend to wander, making us more aware of the passing minutes.

Moreover, our emotions also have a significant impact on our perception of time. Studies have shown that negative emotions such as fear or anxiety can make time feel like it is moving slowly, while positive emotions like joy or excitement can make time feel like it is speeding up. This can explain why time feels so slow when we are waiting for something stressful or unpleasant to occur, but flies by when we are enjoying ourselves.

Our past experiences and memories also play a crucial role in shaping our understanding of time. Our perception of time is intricately linked to our memory as we often use past events as a reference point for how long something took. When we are experiencing something new and unfamiliar, our brains have no point of reference, making it harder for us to gauge the passage of time accurately. This is why our childhood summers seemed to last forever compared to our adult summers, which seem to fly by.

Another fascinating factor that influences our time perception is culture. Diverse cultures have different perceptions of time, with some placing a greater emphasis on the present moment while others prioritize the future or the past. For example, in Western cultures, punctuality is highly valued, and people tend to view time as a finite resource that needs to be used efficiently. In contrast, in many Asian cultures, time is seen as cyclical and fluid, with less emphasis on punctuality and more focus on experiencing the present moment.

Moving on to the neurological factors that affect our understanding of time, recent studies have found that our brain's internal clock is responsible for regulating our perception of time. This internal clock is in an area of the brain called the suprachiasmatic nucleus (SCN), which is responsible for regulating our circadian rhythm (sleep-wake cycle). The SCN sends signals to other parts of the brain that help us track time and make sense of it.

Furthermore, research has also shown that changes in brain chemistry can alter our perception of time. For example, people with attention deficit hyperactivity disorder (ADHD) tend to experience time differently than those without the disorder. They may perceive time as moving faster or slower than it is due to differences in dopamine levels in their brains.

Similarly, individuals who have suffered brain injuries or strokes may also experience changes in their perception of time. This is because damage to specific areas of the brain can disrupt its ability to process and track time accurately. For instance, damage to the prefrontal cortex, the area responsible for executive functions such as planning and decision-making, can lead to time distortions.

Understanding the connection between psychology and time perception can have practical implications in various aspects of our lives. For instance, in education, teachers can use techniques to engage students' attention to make lessons feel shorter. In the workplace, employers can be mindful of employees' emotions and workloads to ensure they do not experience time-related stress or burnout. In healthcare, understanding how time perception is affected by certain disorders can help doctors diagnose and treat patients more effectively.

@@@@

CHAPTER 19

PARADOXES OF TIME

"The Time Paradox reveals how to better use your most irreplaceable resource, based on solid science and timeless wisdom."

—*Martin Seligman*

Time, the most elusive and yet ever-present concept in our lives. It is something we constantly try to grasp, understand, and measure, but it remains a paradoxical mystery. In our daily lives, time seems to pass by quickly or slowly depending on our activities and emotions. We are often chasing it, trying to make the most of it, and yet we feel like we never have enough of it. The paradoxes of time explore this complex and multifaceted concept, shedding light on our perception and understanding of time.

One of the most famous paradoxes of time is the Twin Paradox proposed by *Albert Einstein* in his theory of relativity. It states that if one twin travels at near the speed of light while the other stays on Earth, the traveling twin will experience time dilation, meaning they will age slower than their sibling on Earth. When the traveling twin returns, they would have aged much less than their twin on Earth. This paradox challenges our understanding of time as a constant and universal concept. It shows that time is relative and can be influenced by factors such as speed and gravity.

Another popular paradox is the Grandfather Paradox. It explores the possibility of time travel and its consequences. Imagine if you could go back in time and prevent your grandfather from meeting your grandmother, thus preventing your own birth. This creates a paradox – if you were never born, how could you have gone back in time to prevent your own existence? This paradox highlights the paradoxical nature of changing events in the past and how it could potentially alter our present and future.

The Bootstrap Paradox is another intriguing concept that questions the origin of objects or ideas. It suggests that an object or idea can exist without having a specific origin or creator. For example, if a person travels back in time and gives Beethoven's symphony to Beethoven himself, who then passes it off as his own creation, where did the symphony truly come from? This paradox challenges our understanding of causality and the concept of time as a linear progression.

The Ontological Paradox delves into the question of existence and reality. It suggests that an object or idea can exist without ever having been created. An example of this is the concept of time itself. We perceive time as a constant and universal force, but it is a human construct that does not have a physical form or existence. This paradox forces us to question our perception of reality and how we define existence.

These are just a few examples of the many paradoxes of time that have puzzled scientists, philosophers, and everyday individuals. These paradoxes challenge our understanding of time and its fundamental nature, leading us to question what we think we know about this concept.

One common thread among these paradoxes is the idea of perception. Our perception of time is subjective and can be influenced by several factors such as emotions, experiences, and external stimuli. When we are enjoying ourselves, time seems to fly by, and when we are bored or waiting for something, it seems to drag on. This phenomenon is known as temporal distortion, where our perception of time is distorted by our subjective experiences.

Our perception of time also changes as we age. To us, as children, days seem longer, and summers seem endless. But as we grow older, time seems to speed up, and years seem to pass by in the blink of an eye. This is due to our brain's ability to create new memories and experiences as we age, making each day feel less significant in comparison.

Furthermore, our perception of time also differs across cultures and societies. In some cultures, time is considered a cyclical concept rather than linear, with events repeating themselves in a never-ending cycle. This differs from Western societies where time is seen as a linear progression with a clear beginning and end.

The paradoxes of time not only challenge our perception of time but also raise questions about the nature of reality and existence. They remind us that our understanding of time is limited and there is still much to uncover and explore. These paradoxes also highlight the interconnectedness of time and space, as well as the concept of cause and effect.

In the modern world, where we are constantly chasing after time and trying to be more productive, it is essential to step back and reflect on the paradoxes of time. These paradoxes remind us that time is not just a clock ticking or a calendar page turning; it is a complex and ever-changing concept that shapes our lives in ways we may never fully comprehend. It is okay to take a step back, slow down, and appreciate the present moment instead of constantly chasing after an elusive concept.

@@@@

CHAPTER 21

<u>CONCEPT OF TIME AS AN ILLUSION</u>

"Time is not precious at all, because it is an illusion. What you perceive as precious is not time but the one point that is out of time: the Now. That is precious indeed. The more you are focused on time—past and future—the more you miss the Now, the most precious thing there is."

— *Eckhart Tolle*

Time has always been a fundamental concept in our lives. We are constantly aware of it – keeping track of the minutes, hours, days, and years that pass by. We use it to plan our days, schedule meetings, and set deadlines. But what if I told you that time may not be as real as we think it is? That it is merely an illusion created by our minds? This may sound like a bizarre idea, but the concept of time as an illusion has been a topic of fascination and debate for centuries.

The concept of time as an illusion can be traced back to ancient philosophy. The Greek philosopher, *Parmenides*, argued that time is an illusion and that there is only one eternal present. He believed that everything exists all at once, and the concept of time is just a human construct. Similarly, the Buddhist philosophy also views time as an illusion. The concept of "no-self" in Buddhism suggests that our perception of time is just a product of our constantly changing thoughts and perceptions.

But how do we perceive time?

Our experience of time comes from our brain's ability to process and organize information. Our brains are constantly receiving sensory information from our surroundings, and it processes this information into a coherent order, creating the illusion of time passing. This is known as the "present moment." However, what we perceive as the present moment is just a snapshot of reality. Our brains take this snapshot and create a sense of continuity, making us believe that there is a past and future.

One way to understand this is through the analogy of watching a movie. When we watch a movie, we see a series of still images projected onto the screen in rapid succession, creating the illusion of motion and continuity. Similarly, our brains take in sensory information in fragments and piece them together to create our perception of time.

Moreover, our perception of time is also influenced by our emotions and the context in which we experience it. Have you ever noticed that time seems to fly by when we are having fun, but drags on when we are bored? This is because our emotions and level of engagement can distort our perception of time. For example, when we are engaged in an activity that we find enjoyable, our brains focus on the present moment, making time seem to pass quickly. On the other hand, when we are bored or anxious, our brains tend to wander to the past or future, making time feel slower.

Another factor that contributes to the illusion of time is memory. Our memories play a significant role in how we perceive time. Our brains store memories in a sequential order, creating a sense of linear time. However, our memories are not always accurate and can be influenced by our emotions and biases. This means that our perception of past events may not be an accurate representation of what really happened, further blurring the line between reality and illusion.

But if time is just an illusion created by our minds, then why do we experience it so consistently? The answer lies in the concept of "time perception." Time perception refers to the subjective experience of time passing. While the concept of time may be an illusion, our experience of it is very real. Our bodies have internal clocks that regulate physiological processes, such as sleep cycles and hormone production. These internal clocks help us keep track of time and maintain a sense of rhythm and routine in our daily lives.

Moreover, society has also played a significant role in shaping our perception of time. We have created calendars, clocks, and other systems to measure and organize time. We have divided it into years, months, weeks, days, hours, minutes, and seconds to help us keep track of it more efficiently. This societal construct has become so ingrained in our lives that it is challenging to imagine a world without it.

However, some scientists and philosophers argue that even our most accurate measurements of time, such as atomic clocks, are still relative. Time is not constant; it is relative to the observer's frame of reference. This is demonstrated by Einstein's theory of relativity, which states that time slows down or speeds up depending on the observer's speed and gravitational force. This shows that our perception of time is not an absolute reality but a construct of our minds and the context in which we experience it.

@@@@

CHAPTER 21

CHALLENGING OUR UNDERSTANDING OF REALITY OF TIME

"I have realized that the past and future are real illusions, that they exist in the present, which is what there is and all there is."

— *Alan Wilson Watts*

Time is a fundamental concept that shapes our understanding of reality. We use it to measure the duration and sequence of events, and it serves as the basis for our daily lives. However, recent studies in physics and neuroscience have challenged our traditional understanding of time, claiming that it may be nothing more than an illusion. This idea, known as the "illusion of time," has sparked debate and raised intriguing questions about the nature of reality. In this chapter, we will explore the concept of illusion of time and how it challenges our fundamental understanding of reality.

To understand the illusion of time, we must first define what time really is. In general terms, time is described as a continuous and measurable progression of events from the past to the present and into the future. However, physicists argue that this definition is limited and does not fully capture the essence of time. In fact, Einstein's theory of relativity showed that time is not a universal constant but rather a relative concept that can differ depending on one's frame of reference. This means that time can be perceived differently by different observers based on their relative motion.

Furthermore, neuroscientists have found that our perception of time is influenced by many factors such as emotions, attention, and memory. For example, when we are engaged in an enjoyable activity, time seems to pass quickly, while moments of boredom can make it feel like time is dragging on slowly. Similarly, when we are fully focused on a task, we may lose track of time completely. These findings suggest that our perception of time is not an objective measure but rather a subjective experience.

So why do we perceive time as a linear progression from past to present to future?

This is where the illusion of time comes into play. The concept suggests that our understanding of time as a flow or sequence of events is an illusion created by our brains to make sense of the world. Our brains have evolved to process information in a linear and chronological manner, and time is simply a construct that helps us organize and make sense of our experiences.

One of the main arguments for the illusion of time is based on the idea that there is no real difference between the past, present, and future. In the grand scheme of things, time is a human construct, and the universe exists in a state of perpetual change. The idea that time moves forward is just an illusion created by our limited perception and understanding of reality. This concept challenges our traditional view of time as a linear progression and instead suggests that all moments in time are equally real.

The illusion of time also raises questions about free will and determinism, which have been topics of debate for centuries. If time is an illusion, then does this mean that our future is predetermined? Can we change the course of events? Some argue that if everything in the universe exists simultaneously, then our future is already set in stone. Others believe that our perception of time as an illusion does not negate the concept of free will, but rather it opens new possibilities for how we understand it.

Moreover, the concept of illusion of time has implications for our understanding of consciousness and the self. Our experiences and memories are what shape our sense of self, but if time is an illusion, then how do we define ourselves in relation to the past or future? Additionally, some theories suggest that consciousness may exist outside of time, which challenges our current understanding of how consciousness functions within the brain.

The idea of the illusion of time has also been explored in various spiritual and philosophical beliefs. For example, some Eastern philosophies such as Buddhism and Hinduism view time as an illusion and advocate for living in the present moment. They believe that by letting go of attachments to past or future events, one can achieve a state of enlightenment. This perspective aligns with the concept of illusion of time, as it suggests that our perception of time is a hindrance to fully experiencing the present moment.

@@@@

CHAPTER 22

OUR FASCINATION WITH THE TIME TRAVEL

"Time is not a line but a dimension, like the dimensions of space. If you can bend space, you can bend time also, and if you knew enough and could move faster than light you could travel backward in time and exist in two places at once."

— *Margaret Atwood*

Time travel has been a popular topic in science fiction for decades. From the iconic *DeLorean* in "Back to the Future" to the *Tardis* in "Doctor Who," the idea of being able to travel through time has captivated the imagination of people all over the world. But what if I told you that time travel may not just be a figment of our imagination, but a real possibility? In this chapter, we will explore the concept of time travel and its potential implications.

To understand the possibility of time travel, we first need to define what it means. Time travel is the concept of moving through different points in time, either into the future or past. This can be achieved through various means, such as using advanced technology or natural phenomena.

The most common way of understanding time is through the theory of relativity, put forth by *Albert Einstein* in his famous equation $E=mc^2$. According to this theory, time is not absolute but rather relative to the observer's frame of reference. This means that time can pass at different rates for different individuals depending on their speed and position in space.

Based on this theory, it is possible for time to slow down or speed up for one person while another person experiences it at a normal pace. This phenomenon is known as time dilation and has been seen in experiments involving atomic clocks on airplanes and satellites traveling at high speeds.

So, if we can manipulate time through speed and movement, could we potentially travel through it? The short answer is yes. Scientists have proposed various theoretical methods of time travel based on our current understanding of physics. Let us look at some of these theories.

1. Wormholes:

One way to potentially travel through time is by using wormholes - hypothetical tunnels through space-time that connect two different points in space and time. The concept of wormholes was first introduced in Einstein's theory of relativity, but it is still a highly debated topic in the scientific community.

The idea is that if we can manipulate the fabric of space-time, we could create a wormhole that would allow us to travel through it and emerge at a different point in time. However, the challenges of creating and stabilizing a wormhole are immense, and we are currently far from being able to achieve this feat.

2. Time Dilation:

As mentioned earlier, time dilation is a real phenomenon that has been observed in experiments. However, to experience considerable time travel, one would have to travel at incredibly high speeds. For example, if you were to travel at 99% of the speed of light for one year, you would return to Earth 7 years into the future.

3. Black Holes:

Another proposed method of time travel involves black holes. These are incredibly dense objects in space that have such strong gravitational pull that even light cannot escape them. Inside a black hole, the laws of physics as we know them break down, making it difficult to predict what would happen.

However, some theories suggest that black holes may lead to different points in time and even different universes. If this were true, it would open the possibility of using black holes as a means of time travel. But again, the challenges of surviving such an intense gravitational force are immense.

While these theories show that time travel is theoretically possible, the practicality and feasibility of achieving it are still unknown. The laws of physics as we know them may not be enough to fully understand and manipulate time. And even if we could manipulate time, there are ethical and moral implications to consider.

One of the biggest concerns with time travel is the possibility of altering the past, which could have unforeseen consequences on the present and future. If we were to change a single event in history, it could have a domino effect, altering the course of humanity. The butterfly effect, where a slight change in one system can result in significant changes in another, would be amplified in time travel.

Another concern is the potential exploitation of time travel for personal gain or power. The ability to go back in time and change events could lead to manipulation and abuse of power. It could also create a paradox where one's actions in the past would prevent them from existing in the present.

Despite these concerns, the concept of time travel continues to intrigue and fascinate us. It allows us to imagine what life could be like if we had the ability to visit different eras and witness historical events. It also raises questions about our perception of time and the limitations of our understanding of the universe.

@@@@

CHAPTER 23

POSSIBLE IMPLICATIONS OF TIME TRAVEL ON CONCEPT OF TIME

"When we see the shadow on our images, are we seeing the time 11 minutes ago on Mars? Or are we seeing the time on Mars as observed from Earth now? It is like time travel problems in science fiction. When is now; when was then?"

– *Bill Nye*

Time travel has been a popular topic in science fiction for centuries, captivating our imaginations and sparking endless debates about its possibility. While the idea of traveling through time may seem far-fetched and impossible, recent advancements in physics and technology have raised questions about its plausibility. But, even if time travel were possible, what would be the implications on the concept of time itself? In this chapter, we will investigate the fascinating world of time travel and explore its potential impacts on our understanding of time.

The concept of time has always been a complex and multifaceted one. It is something that affects every aspect of our lives, yet we still struggle to fully understand it. From the ticking of a clock to the aging of our bodies, time is an ever-present force that shapes our reality. But what if we could manipulate time and travel to different points in its continuum? How would it affect our perception of time and its fundamental principles?

One of the most significant implications of time travel is the idea of altering the past. According to the theory of relativity, time is relative, and it can be stretched or compressed depending on factors such as gravity and velocity. This means that if one were to travel close to the

AN INTRODUCTION TO TIME

speed of light, time would slow down for them compared to someone who is stationary. As a result, they could potentially travel to the future by aging at a slower rate than those on Earth. However, traveling to the past poses a more significant challenge.

The concept of changing the past through time travel raises many questions about causality and the possibility of creating paradoxes. If one were to go back in time and alter a past event, it would create a ripple effect that could change the course of history. For example, someone could go back in time and prevent their parents from meeting, thus erasing their own existence. This is known as the grandfather paradox, and it highlights the potential dangers of altering the past.

Furthermore, the idea of changing the past also challenges our understanding of free will. If time travel were possible, it would mean that every event in our lives has already been predetermined and can be altered by someone from the future. This raises questions about the concept of choice and whether our decisions truly have an impact on our future.

Another implication of time travel is the idea of parallel universes or alternate timelines. In this theory, every outcome of an event creates a separate reality, resulting in an infinite number of parallel universes coexisting alongside ours. This means that if one were to travel back in time and change a past event, they would create a new timeline without affecting their own. This concept not only challenges our perception of time but also raises questions about the existence of multiple realities and the consequences of meddling with them.

Moreover, the concept of time travel also has implications on our understanding of the arrow of time. The arrow of time refers to the asymmetry between past and future, as events can only move forward and not backward. However, if one were able to travel through time, this fundamental principle would be challenged. It would also raise questions about the nature of time itself – is it a linear progression or a cyclical loop?

Furthermore, time travel could also have significant implications on our perception of mortality. If one were able to travel through time and potentially live forever, it would blur the lines between life and death. It would also raise ethical concerns about who should have access to such technology and what impact it could have on society.

On a more practical level, time travel could also have implications on historical events and how they are perceived. The ability to go back in time and witness past events firsthand could change our understanding of history and potentially even debunk long-held beliefs. It could also impact how we approach future events, as we would have the ability to see their outcomes and potentially alter them.

However, it is essential to note that the concept of time travel is still highly theoretical, and the technology to achieve it is not yet within our grasp. The laws of physics, as we currently understand them, make it seem almost impossible. But as advancements are made in fields such as quantum mechanics and wormhole theory, the possibility of time travel becomes less of a fantasy and more of a plausible reality.

@@@@

CHAPTER 24

STOPPING TIME

You cannot stop time. You cannot capture light. You can only turn your face up and let it rain down.

- *Kim Edwards*

Time, the very essence of our existence, constantly ticks away, never pausing or slowing down for anyone or anything. It is an inevitable force that governs our lives, from the moment we are born until our last breath. We are all familiar with sayings such as "time waits for no one" and "time is money", emphasizing the importance and urgency associated with time. But what if we could manipulate time? Imagine being able to pause time at will, to extend moments of joy or shorten moments of pain. It is a concept that has fascinated humans for centuries, leading to the question - is it possible to stop time?

The concept of stopping time has been explored in various forms of literature, movies, and even scientific studies. From *H.G. Wells'* "The Time Machine" to iconic movies like "Groundhog Day" and "Click", the idea of controlling time has captured our imagination. However, is it just a mere fantasy or can it be a reality?

To answer this question, we must first understand the concept of time. In simple terms, time is the measure of duration between events. It is a human-caused construct that helps us make sense of our surroundings and organize our lives. But on a deeper level, time is relative and not absolute. It varies depending on factors such as gravity and speed, as proven by Einstein's theory of relativity.

So, if time is relative, does that mean we can manipulate it? The short answer is no. We cannot stop time entirely because it is always moving forward. However, we can slow it down or speed it up to a certain extent. This phenomenon is known as time dilation.

Time dilation occurs due to the difference in perception of time between two observers who are moving at different speeds or in different gravitational fields. This effect was first observed by physicist *Albert Einstein* in his famous thought experiment called the "Twin Paradox". In this experiment, one twin stays on Earth while the other travels in a spaceship at near-light speed. When the traveling twin returns to Earth, they would have aged slower than their Earth-bound sibling. This is because time passes more slowly for objects in motion compared to those at rest.

Similarly, time dilation also occurs due to the influence of gravity. Objects with a higher gravitational pull, such as planets and stars, have a greater effect on time. This phenomenon was proven by the famous *Hafele-Keating experiment*, where atomic clocks were flown around the world in opposite directions and showed a slight difference in time when compared to stationary clocks on Earth.

While these experiments prove that time dilation is a real phenomenon, it is not something that can be controlled or manipulated by humans. It is a natural occurrence governed by the laws of physics and cannot be altered at will.

Another concept often associated with stopping time is freezing time. This refers to the idea of being able to pause time completely, allowing an individual to move freely while everything around them stays motionless. This concept has been explored in science fiction and fantasy genres, but it is not scientifically possible.

The reason for this is that everything in our universe is in constant motion. From the smallest atoms to massive galaxies, everything is continuously vibrating and moving. Even when we think we are standing still, we are moving with the rotation of the Earth. Therefore, it is not possible to freeze time as there would always be some form of movement present.

AN INTRODUCTION TO TIME

Moreover, our perception of time is tied to our consciousness. Our brains process information at a specific rate, which gives us the illusion of time passing at a constant speed. So even if we were somehow able to pause everything around us, our perception of time would continue at its usual pace.

However, advancements in technology have allowed us to simulate the effects of stopping or freezing time through virtual reality and video editing. We can create the illusion of stopping time by slowing down or pausing footage, giving the impression of time standing still. But this is just an illusion and not a true manipulation of time.

In conclusion, while we may not be able to stop or freeze time, we can certainly make the most of it by learning to manage it effectively. Time is a valuable resource, and it is up to us how we choose to use it. We can choose to be consumed by the constant ticking of the clock or learn to live in the present moment, cherishing every second.

So, the next time you find yourself wishing for more hours in a day or trying to stop time from passing, remember that it is a natural and inevitable force that cannot be controlled. Instead, focus on making the most of the time you have and enjoy each moment as it comes. As the saying goes, "time flies when you're having fun", so let us make every moment count.

@@@@

CHAPTER 25
THE EPILOGUE

As the final page of "An Introduction to Time: Universe's Greatest Secrets" turns, the journey through the depths of time and its mysteries ends. It has been a long and complex journey, filled with theories, contradictions, and thought-provoking insights. Now, as we reach the conclusion of this book, it is time to reflect on the concepts and ideas that have been explored and to ponder upon the ultimate understanding of time.

Throughout this book, we have examined the various theories and perspectives on time, from ancient philosophers like Aristotle to modern-day scientists like *Albert Einstein and Stephen Hawking*. We have questioned the nature of time and its existence, pondered upon its relationship with space, and tried to understand its mysterious ways. And yet, despite all our efforts, time remains an enigma – an illusion that continues to baffle us.

As we take a step back and reflect on this journey, one thing becomes clear – time is not just a scientific concept; it is a deeply philosophical and psychological one as well. Our perception of time is shaped by our experiences, emotions, and beliefs. It is a subjective construct that varies from person to person. And this is one of the greatest illusions of time – that it is universal and objective.

The concept of time has evolved over centuries, with diverse cultures and civilizations viewing it in their own unique ways. For some, time is cyclical, while for others, it is linear. Some view it as a river that flows continuously, while others see it as a series of moments frozen in eternity. And yet, all these perspectives are simply different facets of the same mystery – the illusion of time.

AN INTRODUCTION TO TIME

One of the most significant contributions to our understanding of time has been made by *Aristotle*. He believed that time is a measure of change – that it exists because things change. This idea laid the foundation for our modern-day understanding of time as a dimension that can be measured and manipulated. It also introduced the concept of causality – the idea that events are connected in a cause-and-effect relationship, and time is what binds them together.

But as we move forward in this journey, we encounter the paradox of time – the fact that while we experience it as a linear progression, it is also relative and malleable. This paradox is best illustrated by Einstein's theory of relativity, which showed that time is not absolute, but rather depends on the observer's perspective and movement.

Einstein's theories revolutionized our understanding of time, leading to the concept of spacetime – a unified framework that combines space and time as one entity. It also introduced the idea of time dilation – the slowing down of time in the presence of massive objects or at high speeds. This concept has been proven through experiments such as the famous *Hafele-Keating* experiment, which showed that time does indeed slow down for objects in motion.

But the most intriguing aspect of *Einstein*'s theory is the concept of time travel. While it is still a topic of science fiction, his equations do suggest the possibility of traveling through time, albeit only in a theoretical sense. The idea of being able to visit different moments in time brings up questions about free will and destiny – if time is fixed, can we change our fate by traveling back or forth in time?

As we continue to grapple with these questions, it becomes clear that our understanding of time is constantly evolving. Every new theory and discovery add a new layer to this complex puzzle. And that is what makes time so fascinating – its ability to keep us questioning and seeking answers.

However, despite all our advancements in science and technology, there are still many aspects of time that we cannot fully comprehend.

For instance, we still do not know what caused the Big Bang or how time began. We also struggle to understand the concept of eternity – a never-ending, timeless existence that goes beyond our mortal understanding of time.

But the greatest lesson that we can take away from this journey through time is that it is futile to try and capture it in a single definition or concept. As humans, we are limited by our perceptions and understanding, and time is a concept that surpasses these limitations. Instead of trying to unravel its mysteries, let us embrace the illusion of time and the wonder it brings to our lives.

In conclusion, "An Introduction to Time: Universe's Greatest Secrets" has taken us on a thought-provoking and mind-bending journey through the depths of time. It has shown us that while we may never fully comprehend its nature, the pursuit of understanding is what keeps us moving forward. And just like time, this book will remain an enigmatic and timeless reminder of the illusion that we call time.

@@@@

CHAPTER 26

FURTHER READING

"If you see an antimatter version of yourself running towards you, think twice before embracing."

— *J. Richard Gott, Time Travel in Einstein's Universe: The Physical Possibilities of Travel Through Time*

Books have always been a source of knowledge, escapism, and entertainment for people. They allow us to explore different worlds, learn about new perspectives, and expand our understanding of the world. One of the most fascinating and overwhelming topics that have captivated readers for centuries is time, space, and the universe. From ancient civilizations to modern-day scientists, the concept of time and space has been studied, debated, and written about extensively. In this chapter, we will discuss some of the most thought-provoking books available on time, space, and the universe.

1. **"A Brief History of Time"** by *Stephen Hawking*: This book needs no introduction. Considered a masterpiece in the field of astrophysics, Stephen Hawking's "A Brief History of Time" provides a comprehensive understanding of the origins of the universe and its future. Hawking explains complex theories like the Big Bang and Black Holes in simple language, making it accessible to readers from all backgrounds. This book is a must-read for anyone looking to understand the basics of time, space, and the universe.

2. **"Cosmos"** by *Carl Sagan*: This book is a journey through space and time, as seen through the eyes of renowned astronomer *Carl Sagan*. With his eloquent writing style and vivid imagination, Sagan takes readers on a journey through the vastness of our universe, exploring everything from the birth and death of stars to the search for extraterrestrial life. The book also delves into philosophical questions

about our place in the universe, making it an enlightening read for anyone interested in science and philosophy.

3. **"The Fabric of the Cosmos"** by *Brian Greene*: In this book, theoretical physicist *Brian Greene* explores some of the most intriguing concepts in physics that shape our understanding of the universe. From quantum mechanics to string theory, *Greene* explains these complex topics in an engaging and accessible way. He also discusses the concept of time and how it is intertwined with space, taking readers on a mind-bending journey through the wonders of the cosmos.

4. **"A Universe from Nothing"** by *Lawrence Krauss*: In this book, physicist *Lawrence Krauss* discusses one of the biggest mysteries of the universe – how did the universe come into existence? *Krauss* explains how our current understanding of the laws of physics suggests that the universe may have originated from nothing. He also explores the role of time in shaping the universe and how it may eventually lead to its demise. This book is a must-read for anyone interested in the origins and fate of our universe.

5. **"The Elegant Universe"** by *Brian Greene*: Another gem by *Brian Greene*, it explores the fundamental principles of modern physics – relativity and quantum mechanics. *Greene's* writing makes these complex concepts easy to grasp, even for those without a scientific background. He also delves into the theory of everything – a single unifying theory that explains all physical phenomena in the universe. This book is a fascinating read for anyone looking to understand the workings of our universe at a deeper level.

6. **"Time Reborn"** by *Lee Smolin*: In this thought-provoking book, physicist *Lee Smolin* challenges our current understanding of time and argues that it is not an illusion but a fundamental aspect of reality. Smolin argues that time is not a mere construct to measure change, but something that plays an active role in shaping our universe. He also discusses how this new perspective on time can help resolve some

long-standing mysteries in physics, such as the unification of relativity and quantum mechanics.

7. **"The Order of Time"** by *Carlo Rovelli*: In this intriguing book, physicist *Carlo Rovelli* challenges our conventional notion of time as a linear progression and argues that it is more complex and elusive than we think. *Rovelli* takes readers on a journey through the history of time, from the ancient Greeks to modern-day physics. He also explores how our perception of time is shaped by our subjective experience and how it is intertwined with our understanding of space and the universe.

8. **"Black Holes and Time Warps"** by *Kip Thorne*: In this book, renowned physicist *Kip Thorne* explores the mysteries of black holes – one of the most enigmatic and fascinating objects in the universe. *Thorne* explains how these cosmic objects deform space and time, creating wormholes that could potentially allow us to travel through time. He also discusses how black holes could be a key to unravelling some of the biggest mysteries of our universe, such as the existence of parallel universes.

9. **"The End of Time"** by *Julian Barbour*: In this thought-provoking book, physicist *Julian Barbour* challenges our conventional understanding of time as a constant and argues that it is an illusion created by our minds. He presents a revolutionary theory called "shape dynamics" that suggests that time is an emergent property of the universe, rather than a fundamental aspect. *Barbour*'s ideas may challenge some deeply ingrained beliefs about time and the universe, making this book a stimulating read for anyone interested in this topic.

10. **"Einstein's Dreams"** by *Alan Lightman*: This book is a work of fiction, but it offers a unique perspective on the concept of time. The author, *Alan Lightman*, imagines different worlds based on different theories of time proposed by *Albert Einstein*. Each chapter explores a different concept, such as time running in reverse or standing still, allowing readers to question their own perception of time and its impact on our lives.

These are just a few of the many books available on time, space, and the universe. Each one offers a unique perspective and delves into complex concepts that challenge our understanding of the world. So, if you are looking to expand your knowledge, explore different theories, and be amazed by the wonders of the universe, pick up one of these books and embark on a journey through time and space.

Here is a comprehensive list of books (including some already described above), documentaries, online resources, and more on time, the universe's greatest secret:

Books:
1. "A Brief History of Time" by Stephen Hawking
2. "The Fabric of the Cosmos" by Brian Greene
3. "Time Reborn" by *Lee Smolin*
4. "The Time Book" by *Clifford A. Pickover*
5. "Warped Passages" by *Lisa Randall*
6. "The Elegant Universe" by *Brian Greene*
7. "The Hidden Reality" by *Brian Greene*
8. "Time and the Universe" by *Sean Carroll*
9. "From Eternity to Here" by *Sean Carroll*
10. "The Big Picture" by *Sean Carroll*

Documentaries:
1. "The Fabric of the Cosmos" (PBS documentary series)
2. "The Universe" (History Channel documentary series)
3. "Cosmos: A Spacetime Odyssey" (FOX/National Geographic documentary series)
4. "Time" (BBC documentary series)
5. "The Story of Time" (BBC documentary)
6. "How the Universe Works" (Discovery Channel documentary series)
7. "Space Time" (PBS Space Time documentary series)
8. "The Secrets of Time" (Smithsonian Channel documentary)

Online Resources:

Websites:
1. NASA's Time and Space website
2. (link unavailable)'s Time and Relativity section
3. Scientific American's Time and Space section
4. (link unavailable)'s Time and Space section
5. The European Organization for Nuclear Research (CERN) website

Podcasts:
1. "The Astronomy Podcast"
2. "The Physics Podcast"
3. "The Time Podcast"
4. "The Universe Podcast"
5. "StarTalk Radio"

Online Courses:
1. Coursera's "Time and Relativity" course
2. edX's "The Universe and Space" course
3. Udemy's "Time and the Universe" course
4. Khan Academy's "Physics: Time and Relativity" course

Experts and Researchers:
1. *Stephen Hawking* (Theoretical Physicist and Cosmologist) (Died 14 March 2018)
2. *Brian Greene* (Theoretical Physicist and Mathematician)
3. *Sean Carroll* (Theoretical Physicist and Cosmologist)
4. *Lisa Randall* (Theoretical Physicist)
5. *Neil deGrasse Tyson* (Astrophysicist and Science Communicator)

Conferences and Events:
1. The International Conference on Time and Relativity
2. The Annual Meeting of the American Physical Society
3. The Conference on the Physics of Time
4. The Symposium on Time and the Universe

Mobile Apps:

1. NASA App (iOS and Android)
2. Space Time (iOS and Android)
3. Time and Space (iOS and Android)
4. Physics Lab (iOS and Android)
5. Astronomy Lab (iOS and Android)

Videos:

YouTube Channels:
1. Crash Course Physics
2. 3Blue1Brown (Physics Animations)
3. PBS Space Time
4. Physics Girl
5. TED-Ed (Physics and Astronomy)

@@@@

CHAPTER 27

GLOSSARY OF TIME -RELATED TERMS

Here's an short glossary of terms related to time:

General Terms
 1. Time: A measure of the duration between events.
 2. Past: Events that have occurred before the present.
 3. Present: The current moment.
 4. Future: Events that will occur after the present.
 5. Timeline: A linear representation of events in chronological order.
 6. Timeframe: A specific period or duration.
 7. Timeless: Unaffected by time.

Scientific Terms
 1. Time Dilation (Relativity): The phenomenon where time appears to slow down or speed up depending on relative motion.
 2. Spacetime (Relativity): The fabric combining space and time.
 3. Chronon (Physics): A hypothetical quantum of time.
 4. Time Reversal (Physics): The hypothetical process of reversing the flow of time.
 5. Entropy (Thermodynamics): A measure of disorder or randomness.
 6. Quantum Decoherence (Physics): Loss of quantum coherence over time.
 7. Time Translation Symmetry (Physics): The concept that physical laws remain unchanged over time.

Astronomical Terms
 1. Cosmic Time: The time measured from the Big Bang.
 2. Galactic Year: The time taken by the Sun to orbit the Milky Way.

3. Sidereal Time: Time measured relative to the stars.
4. Solar Time: Time measured relative to the Sun.
5. Astronomical Unit (AU): A unit of time and distance.
6. Epoch (Astronomy): A reference point in time.

Philosophical Terms

1. Eternalism: The belief that past, present, and future exist simultaneously.
2. Presentism: The belief that only the present exists.
3. Block Universe Theory: The concept of spacetime as a fixed, unchanging block.
4. Timelessness: The idea that time has no objective existence.
5. Determinism: The concept that events are predetermined.

Physics Terms

1. Time Constant (Physics): The time required for a system to respond to changes.
2. Period (Physics): The time taken for a cyclical process.
3. Frequency (Physics): The number of cycles per unit time.
4. Oscillation (Physics): Repeating patterns over time.
5. Wave Function (Physics): A mathematical description of quantum systems.

Theoretical Terms

1. Wormhole (Theoretical Physics): A hypothetical shortcut through spacetime.
2. Time Travel (Theoretical Physics): Moving through time.
3. Quantum Fluctuation (Theoretical Physics): Random changes in energy.
4. Black Hole Information Paradox (Theoretical Physics): The puzzle of information loss in black holes.

Other Terms

1. Circadian Rhythm (Biology): The internal biological clock.

2. Time Management (Psychology): Controlling time usage.
3. Time Perception (Psychology): The subjective experience of time.
4. Time Estimation (Psychology): The ability to estimate time intervals.

This glossary covers some aspects of time, providing a preliminary foundation for understanding its complexities.

"An Introduction to Time: Universe's Greatest Secret" by Dr. A.K.Saxena, Ph.D., November 2024, New Delhi, India

@@@@

Don't miss out!

Visit the website below and you can sign up to receive emails whenever Dr.A.K.Saxena, Ph.D. publishes a new book. There's no charge and no obligation.

https://books2read.com/r/B-A-GTBRC-ESSHF

BOOKS2READ

Connecting independent readers to independent writers.

Did you love *An Introduction to Time*? Then you should read *The "Karma" Puzzle* by Dr.A.K.Saxena, Ph.D.!

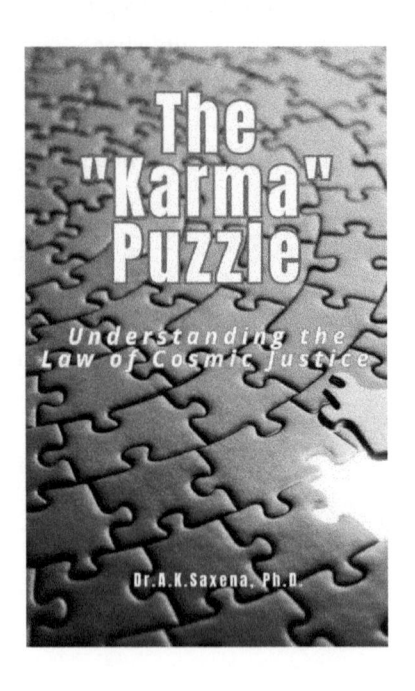

"The "Karma" Puzzle": Understanding the Law of Cosmic Justice" is a profound exploration of the ancient and universal principle of *"Karma"*, a concept that has captivated human imagination for centuries. This book delves into the intricate web of cause and effect, revealing the underlying mechanisms that govern our lives and shape our destinies.

For millennia, philosophers, spiritual leaders, and wisdom seekers have grappled with the enigma of *"Karma"*. From the sacred texts of Hinduism and Buddhism to the philosophical musings of ancient Greece and modern thought leaders, the concept of *"Karma"* has evolved, yet its essence remains unchanged: every action, thought, and intention has consequences that reverberate throughout the cosmos.

As we navigate the complexities of life, we often find ourselves pondering fundamental questions:
Why do good things happen to bad people?
Why do we suffer despite our best intentions?
How can we break free from patterns of pain and stagnation?
The *"Karma"* offers answers, providing a comprehensive framework for understanding the cosmic justice that governs our existence.

This book is not merely a theoretical exploration but a practical guide for living in harmony with the karmic principles that govern our universe. Join me on this transformative journey as we unravel the mysteries of the *"Karma"*.

"The "Karma" Puzzle" is more than a book – it's a key to unlocking the secrets of the universe and your place within it. As we embark on this exploration, may the ancient wisdom of *"Karma"* illuminate our path, guiding us toward a life of purpose, balance, and harmony.

Also by Dr.A.K.Saxena, Ph.D.

Handbook for Introverted Leaders: Strategies for Success
No Longer a Yes Man
Fasting, Feasting and Spirituality
The "Karma" Puzzle
Open Secrets
15 Things You Should not Worry About
Why Denial of Death?
From Sorrow to Serenity
Unlock Your Inner Power and Potential
An Introduction to Time

About the Author

Dr.A.K.Saxena is a former bureaucrat in Government of India, New Delhi, India. He holds a Doctorate degree in science from prestigious University of Lucknow, is a qualified and practicing lawyer in Supreme Court of India. Dr. Saxena is also a motivational speaker, life coach and a prolific writer. He writes mostly on self-help, parapsychology and spirituality. Besides writing, his interests are reading science-fiction, Urdu poetry, astrology, parapsychology, metaphysics, mysticism, spirituality, Artificial Intelligence and Machine Learning. Dr. Saxena loves cooking and is fond of Indian vegetarian cuisine. His beautiful wife Vijay Shree has been Indian National Award Winner Teacher of Hindi and Sanskrit languages. She writes Hindi poetry and is a great storyteller too. They have two little lovely granddaughters.